Owned by an Eagle

GERALD SUMMERS

Owned by an Eagle

drawings by Eva Hülsmann

E. P. DUTTON

NEW YORK

This book is dedicated to Imogen and my 'instant' family in gratitude for their encouragement and forbearance; to my friend Clifford Laurie without whose help and hours of unrequited labour it would never have been written or interpreted into anything approaching legibility; and finally to Random in the hope that her next sixteen years will prove as eventful as the last.

Copyright © 1976 by Gerald Summers
All rights reserved. Printed in the U.S.A.
First American Edition

Library of Congress Catalog Number: 76-52316
ISBN: 0-525-17434-6

10 9 8 7 6 5 4 3 2 1

Prologue

High above Llyn y Fan, black against the wintry sky, an immense bird is cruising in airy circles, now with motionless wings outstretched, each great primary feather feeling for the up-currents, as if resting on her oars, now smiting the un-resisting air as if she owed it a personal grudge, thus gaining impetus for further gyrations.

The casual observer, glancing skywards, confused by distance, might be forgiven for mistaking this bird for yet another buzzard, for this is buzzard country. It is too the haunt, indeed the last British refuge, of the red fork-tailed kite, but this bird, now outlined against the mountain slopes glinting with the first snow of winter, is neither kite nor buzzard. The lazy arrogance of her flight has a contemptuous ease about it, a finish which is lacking from the buzzard's moth-like drifting. As I watch, she seems to trim her sails and, with one long slanting glide, is overhead in seconds. She pitches in the top branches of the oak tree under which I am standing and at once greets me with her customary throaty warble, a voice unexpected in so powerful a bird.

She is a golden eagle named Random and she has been my constant companion by day, and not infrequently by night, for close on sixteen years. This is the story of our friendship, a friendship which for many years was shared with my saluki, Tally, with whose release from quarantine, after my return from Kenya, this book opens.

Chapter One

I put down the letter I had just read with such fumbling
excitement and walked out through the french windows on to
the lawn outside our house at Sutton Valence in Kent. The
house stood on the slopes of a ridge which faced southwards
to the mist-shrouded Weald, now crisp and golden with the
oncoming of another autumn. The view was such as I had often
dreamed about when beneath the steely blue skies of distant
Africa, soft and gently welcoming as only the British countryside can be; but now I hardly noticed it. The letter had been
short and to the point, but the few lines of impersonal type-
script contained all that I could wish to read.

Tally was coming home at last! Tally, the saluki puppy,
who for a few brief months had been my well-loved companion in Kenya, had finished his quarantine and was to be
released after six long heart-searing months of separation. I
had known, of course, that the day of our reunion was fast
approaching but here was official confirmation.

On the previous evening an ornithologist had called and
persuaded me (as if I needed persuasion!) to take on the role
of foster-father to an orphan golden eaglet. Pleasures and
responsibilities, it seemed, came not singly but by battalions.

It appeared that my caller, an enthusiastic ornithologist,
had just returned from a visit to Spain. His intention had been
to study and, if possible, to photograph the azure-winged
magpie, an exotic-looking member of the crow family, whose
European range is restricted to the Iberian peninsula. In the
course of his wanderings he had reached the High Sierras,
where, slaking his thirst one evening in a wayside hostelry, he
had heard news of a pair of golden eagles which were breeding

7

on a cliff face at no great distance from where he was camping. It seemed that the eagles, who were rearing two well-grown youngsters, had been condemned, probably unjustly, for killing young ibex kids. In those days, whatever the position may be today, eagles were not protected in Spain. In any case ibex were much rarer than golden eagles, which, in various sub-species, are distributed all round the northern hemisphere, ranging in size from the huge, wolf-killing Russian Berkute to the comparatively smaller, insignificant variety found in Japan.

The ornithologist's informant was, as it happened, the Spanish equivalent of a game warden who had received orders to destroy the eagles without delay. The Englishman was understandably horrified and, after an impassioned plea, a compromise was reached. It was agreed that, if the young eagles were removed from the eyrie, the old birds would be spared and, by the firing of judicious shots across their bows, encouraged to make their headquarters elsewhere. The following day the intrepid ornithologist, aided by his Spanish companion, descended the cliff on a rope, stuffed the two eaglets, protesting violently, into his rucksack and, after many vicissitudes, arrived at his flat near Sloane Square. Hearing through the ornithological grapevine of my obsession with anything resembling a bird of prey and considering a London flat, however luxurious, an unsuitable substitute for an eyrie, he lost no time in tracking me down.

I dressed in the blustery autumn dawn. High on the pinnacle of the solitary cypress tree, our resident mistle-thrush, swaying like a tiny animated weather-vane, was shouting out an especially triumphant paean of welcome. Tally was coming home today! I snatched up a cup of strong tea and groped for a hunk of buttered toast, which I somehow managed to stow away before slipping out into the grey drizzly half-light to catch the bus to Maidstone and the train. The kennels opened at 10 a.m. and I was going to be there on the dot. Slowly, maddeningly slowly, the train clanked and ground its

8

serpentine way across the sodden Kentish countryside, pausing at little-used stations where few entered or alighted. Somehow we came at last to Woolwich, noisy with the impersonal rush of city commuters. I looked at them with pity. There was no Tally waiting with ears cocked for their approaching footsteps.

A short bus-ride and I was there, striding through the newly-opened gates, trying to conceal the pulsing excitement I was feeling. The kennel-maid took me to the kennel, where, with feathery ears half-raised and plumed tail gently waving, Tally awaited me. We had not seen each other for more than three months but there was no hesitation or hint of coolness in his welcome. Salukis are not given to showing their feelings and Tally was as reserved as any, but on this occasion he did

something he had never done before and was never to repeat. As I slipped into the narrow, barred kennel he reared up, putting his great tawny paws on my chest and, throwing back his delicate gazelle-like head, gave forth a long yodelling cry, half whimper, half bark, a cry of sheer pent-up emotion released at last after six months of hope and resignation. I was amazed at his appearance, having forgotten just how handsome he was. He had grown until he was nearly as tall as a deer-hound (he was later acknowledged to be one of the biggest of his breed in Britain). His condition was superb. His coat gleamed, in parts blacker than night itself, in parts redder than the mahogany-red dust of his native desert. He had put on weight, but that was to be expected. His whole appearance reflected the care he had received. I snapped on the lead I had brought and, avoiding the reproachful gaze of those whose day of liberation was yet to come, strode out through the gates with Tally dancing beside me.

I decided to walk to the station to give him a chance to get to know me again and, more important, to realize that this time we were never to be separated. Tally paced beside me with his light, springy, almost feline desert trot. His eyes were everywhere, alert and nervous as a long-stabled Arab stallion. It was raining relentlessly and I had forgotten my mackintosh but it could have been May at its balmiest for all I cared. At the bus-stop a dripping queue were waiting. At the sight of Tally their eyes lit up and they made admiring noises, some daringly reaching out to pat his silky head as he pranced laughing by. On the opposite side of the road a young couple were walking with a dog – a big, wolfish-looking sable alsatian. Tally saw the dog and without warning lunged forward, all but jerking the tough, plaited lead from my grip. Passing a newsagent's shop, a jolly, hairy, black mongrel with a ridiculous tail like an animated question-mark and a face like the 'before' bit of a shaving-cream advertisement, trotted confidently up to inspect this exotic stranger. Tally's hackles rose slightly and his tail stiffened aggressively. I spoke to him, and after exchanging sniffs, he relaxed and even showed some signs of wanting to gambol with this disreputable-looking cockney. I found out later that he had a deeply-implanted dislike and distrust of alsatians and alsatian-like dogs, including elk-hounds, which he made no attempt to conceal. He didn't care much for other dogs either, but unless they turned tail and fled at the sight of him, as often happened, he generally ignored them until he came to know them better, when he might even become friendly in a rather condescending way.

A small, dejected-looking gaggle of sodden passengers clustered round the bookstall at the station. A few of them spoke to Tally and, as an afterthought, to me. In those days, salukis (and indeed other exotic breeds) were far less common in Britain than they are today. In fact, few people had ever seen one or even heard of one. Things unfortunately are differ-ent today, when a measles-like rash of what can only be described as puppy-breeding factories has arisen over the face of the country to cater for the status-seekers, who, in many cases, when they discover that a St Bernard puppy grows larger than a cocker spaniel, either pass it on to an even more unsuitable home or, worse still, turn it adrift to seek its fortune as best it may. In the empty compartment Tally curled beside me on the seat with his head on my knee, just as he had done on our first journey together from Nairobi to Manyani, a journey that seemed a lifetime, indeed a whole world away.

My mother had never met Tally but had heard so much about him that she felt she knew him intimately. Their meeting was just as one would have expected from two such natural aristocrats. Each extended a gentle courtesy towards the other. Tally walked forward slowly, his tail gently swaying and a smile of friendship in his eyes; my mother momentarily rested the fingers of one hand upon the top of his silky black head before stepping aside to let him walk past her into the welcoming, lamplit hall. That night, tired but happy, he lay, a great black shadow, curled upon a rug beside my bed, his quiet breathing punctuated at long intervals by deep sighs of

pure well-being. He was utterly content to have come home at
long last to the place where he belonged.

In spite of his long incarceration, he was remarkably fit,
bodily and mentally, and he appeared to suffer no neuroses.
His calm, philosophic temper was unchanged. There was,
however, one problem – a big one. Not only did he chase other
dogs on sight, he chased everything that moved, and was
constantly on the alert for signs of potential quarry. I could
see that, living in a densely-populated area of southern
England, we were in for trouble unless something could be
done and done quickly.

The problem of other dogs was the first to bother us. It so
happened that some friends of my mother had a large, tough,
white bull-terrier. Like most bull-terriers, Odo was fearless,
but unlike the majority of his breed he was unaggressive and
much preferred playing to fighting. He was about the same
age as Tally, still virtually a puppy. When Odo came to call
Tally was at the far end of the garden, sniffing at a hole that
had long before been occupied by a solitary cony, now defunct,

slain as a garden raider and later eaten because there was still a shortage of meat. Odo, a friend of all the world, came lumbering up, his tail erect, the tip vibrating like a compass-needle gone mad and grinning as only a bull-terrier can grin, the piratical black patch over one eye belying the guilelessness of his sunny good nature. Tally, hearing his heavy, earth-shaking tread, looked up and immediately forgot the challenge of the non-existent rabbit. With ears and tail proudly raised and his hackles looking like the cloak of a Masai warrior he forth-with fell upon the intruder and grabbed him by the neck. His powerful jaws made little impression on the bull-terrier's gutta-percha skin, and Odo, obviously tickled pink, shook him off and went rampaging round the lawn, overjoyed at the prospect of a romp with this splendid new acquaintance.

This was an invitation even the haughty Tally could scarcely refuse and in seconds they were ploughing through the autumnal flower-beds, leaving a swathe of destruction in their wake and my mother inarticulate with justifiable, though frustrated, rage. However, it did the trick. From that time on, provided a dog was willing to be friendly and did not upset his rather exaggerated sense of dignity, Tally was prepared to accept the friendship offered. He was, however, a fierce and terrible fighter when the situation justified it. With his speed of movement, weight, and exceptional strength of jaw, he could hold his own with practically anything that was rash enough to provoke him, until, that is, the day he came up against Random, the golden eagle.

The day after his arrival I took Tally to the downs above Lewes, the downs I knew so well and had missed so deeply. Here, high above the grey-green channel, where the rose-breasted linnets fluttered among the golden-flecked gorse bushes, and the yellow-hammer, spruce in rust-brown and primrose livery, endlessly pleaded for 'little bits of bread and no cheese', I gave him his head. Feeling the elastic, thyme-scented turf beneath his feet, he reacted like the highly-strung thoroughbred that he was, floating over the chalky, sheep-cropped turf with all the devastating speed and rhythmic grace of an earth-bound falcon. It was plain that captivity had done little to slow him up.

Tally was ranging at a hand-gallop on my right flank when I all but fell over a big jack hare which had been dozing in his shallow form and digesting his latest meal of turnips purloined from a farm in the valley far below. The hare was off, breasting the slope ahead with a few pile-driver thrusts of his immensely powerful hind legs. Tally must have been a good three hundred yards below the quarry when my shout drew his attention to what was afoot. Three sky-hopping bounds he gave to get the hare well and truly in his sights and then he was off at a pace that could make even the fleet African gazelles look to their laurels. The hare was tough, wily, and he knew his ground well. He knew where he was going and he meant to get there with as little delay as possible. There are few speedier animals than a fit downland hare at the peak of his condition. The gap dwindled none the less, as the hare strove all out for his objective, a patch of impenetrable gorse a quarter of a mile ahead.

This was coursing at its best — one hound against a fit, determined quarry, with all the odds in favour of the hunted. Tally was now within a length of the prey and to anyone who didn't know hares the outcome looked inevitable. However, I did know hares and I knew just what to expect. One second the fugitive was heading flat out for the sanctuary, now drawing ever closer, the next he was going equally fast in the opposite direction. As the saluki drove in for the kill the hare, timing it to the smallest fraction of a second, had jinked, turning completely in his own length, without losing an iota of his momentum. Tally, thrown out completely, swung round in a wide arc and came in again more determined than before. The hare was now on course again and the refuge he sought was within yards as Tally once more closed the gap, his back and tail lying in one tense line of concentration as he swept in from the other flank, his jaws already aiming straight for his quarry's back. Hare and hound seemed momentarily to become one as the old jack, his wits never failing him, swerved once more when all seemed over. Somehow, by a miracle of good timing, he emerged unscathed to disappear into the welcoming citadel of spiky, close-growing bushes.

Tally, it was obvious, would have to learn a great deal more about hares than that. Seeing his erstwhile victim disappearing, he eased his terrific pace slightly and loped round to the far side of the thicket in the hope, apparently, that the hare would be sporting enough to emerge once more for a second round of this well-matched contest. Luckily the hare had more sense. In any case, Tally had had enough. He might not have lost the edge of his speed but it was distressingly clear that his fitness and stamina left much to be desired. When he joined me, as I lay on my back on the springy, sweet-scented turf, revelling in the thrill of the chase, his flanks were heaving and his heart throbbing like a distant motor-cycle engine. We lay resting together, listening to the tinkling song of invisible sky-larks and the distant bleating of sheep in their fold behind a shoulder of the downs. Tally, his wind regained, drank his fill from a dew-pond, one of those mysterious circles of pure water of which no one seems to know the true origin. Together we dropped down into the valley as the lights were appearing in the little downland villages beneath us, until at last, stiff, empty-handed, but content in our companionship, we caught the bus for home.

That night, as I lay in bed, listening to Tally's gentle breathing and the occasional heavy thump as he shifted his position, I thought deeply. Indeed I had much to ponder upon. The next day my golden eagle eyas was due to arrive. To what, I wondered, had I committed myself now? Only one British falconer, to my knowledge, had successfully trained a golden eagle up to that time, and that eagle was a male — the world-famous Mr Ramshaw, lifelong companion of Captain Knight who was about the only practising falconer that I had ever met. True, Knight had, prior to owning Ramshaw, succeeded

16

to some extent with two Scottish female eagles, both named Grampian; but they had proved so unpredictable and aggressive that they had had to be released in the land of their origin. The only books I had read on the subject gave the sort of much-quoted advice that Mr Punch is supposed to have given to those about to be married: Don't! I wondered how many falconers had attempted what I was about to attempt, and how many of these had persevered. This was the sort of challenge that really appealed to me. Could I succeed where others, with far more experience, had failed? I would only too soon find out. I had nothing to go on — there was no manual on the training of golden eagles, at least none that I knew about. I would have to play it largely by ear.

Chapter Two

The next day Random arrived, travelling unceremoniously in the boot of a battered saloon. It was a prosaic container, but as I stood gazing at it from the outside it needed little imagination to tell me what it must contain. After some frustratingly prolonged fumbling with the handle, the driver opened the boot, and there before me, in all her appealing helplessness, lay the being who has probably had more influence on my life than any other non-human creature. Her watchful, imperious head, still adorned with snowy-white flecks of baby down, was turned towards me, and her huge, deep-set, hazel eyes looked into mine with an expression of fearless trust which went right to my heart as I gathered her into my arms and placed her on the grass. Here on the sunny lawn we sat side by side; she, alone, defenceless, and yet completely mistress of the situation. So must the infant Boadicea, Queen of the Iceni, have looked at a comparable age!

There is something particularly disarming about the young of predatory birds and beasts. The gawky, out-at-elbows look of the cubs of the tiger and leopard gives little indication of the lethal grace and dexterity that will eventually be theirs. So it was with Random. Her great feet, with their annihilating black talons, were the colour of a newly-opened king-cup, as was the cere above her curved, steely, blue-black beak. Her body plumage, what there was of it, was a subtle blend of milk and plain chocolate and there was, as yet, no trace of the metallic, bronzy mantle which was later to flow down from her crown to her scapulas in the glorious cascade of colour which gives her species its popular name. Her tail, all that could be seen, was a creamy white colour set off by a broad, dark terminal band. This white tail she was to keep for the first three or four years of her life. It was to lead many visitors into suspecting her of being a sea eagle, although, paradoxically, the juveniles of that clumsy-looking raptor have a dark tail until maturity is reached.

As soon as she was comfortably settled on the lawn and had taken a leisurely and somewhat haughty look round at her new surroundings she shook herself like a newly unsaddled horse, 'rousing' to show her confidence in herself and in her new acquaintances. Her crop was still half distended but she opened her enormous gape and yelped loudly, indicating that she wanted something and wanted it immediately. She was obviously hot and probably thirsty after her unceremonious travelling arrangements, so I filled a shallow canvas basin which I had used during my travels in East Africa. This was filled to the brim. Random shuffled forward on folded tarsi, which were obviously unable to support her considerable weight. I lifted her into the shallow water and she began to drink immediately. Lowering her great rapacious beak into the water, she spooned it up a few drops at a time, throwing back her head to let it trickle down her parched throat, her clear grey eyes half shut with the rapture of the moment. Watching her, I reflected on moments of aching thirst which I had suffered in East Africa and I savoured every drop with her as she drank.

Her thirst quenched, she began to bathe, wallowing and heaving like a porpoise, her half-developed wings thrashing the surface as she ducked and bobbed like a pigeon in a puddle, until the water's surface was flecked with white fluff, like floating thistledown, and her baby plumage was soaked to the roots of her soft grey-shafted blood quills. I lifted her out and put her gently on the lawn once more, where she shook herself, spattering droplets of silver water far around her. Reaching out her beak she tweaked at a plantain stem, snipping it off close to the ground, tossed it over her back and looked round for something else to amuse her, for she was as playful as a half-grown puppy. I had prepared her meal and she took it from my fingers, wolfing it down with satisfactory greed, yelping querulously if I was slow in producing the next bit. She ate until she looked as if she had swallowed an outsize tennis ball, and it was I, not Random, who decided that she had had enough.

At this point Tally stalked on to the scene, sauntering across the lawn with the peculiar controlled slouch which is so typical of the saluki, who is in no hurry, is in full command of himself and his surroundings and doesn't give a damn who knows it. He strolled up to Random, who raised her hackles slightly at his approach. Tally reached out a long aristocratic muzzle to sniff the curious object before him. Random's hackles rose higher still and a hard, unfathomable look came into her deep grey eyes. Tally, ignoring these signs, gave her a watchful glance, picked up a piece of meat lying a few feet from Random and swallowed it without further sign of interest, moving away with his springy, daisy-cutting trot, head and plume tail held high. He didn't realize it at the time, but he had made a dangerous enemy.

I had prepared an amateurish sort of imitation eyrie in a well-lit out-house, and here I would keep her, knowing that it would be some weeks before I could begin training her. Pulling on a clumsy, but I hoped eagle-proof, glove, I pressed it just below her distended crop and somehow, with a gentle push, managed to hoist her on to my fist. Her already impressive wings flapped wildly. Her feet, which had but little power at that time, somehow found a hold and for the first time she was standing on my arm, gazing round her with complete composure. My arm sagged under her weight, but it was one of the proudest moments of my life. Anxious to avoid anything approaching an anti-climax I hurried her to her new home and settled her on her nest, leaving her to rest and digest her gargantuan feast in peace.

I took Tally for a long walk in the woods that still clustered darkly in the Kentish Weald below our house. We both enjoyed ourselves, Tally bounding after rabbits that flitted between the fern brakes or leaping almost into the lowest branches of the trees in pursuit of the ubiquitous but elusive grey squirrels. Some day, Tally always hoped, one of these would somehow make a mistake, a mistake which he would make sure was never repeated, at least by that particular squirrel!

I loved these walks and I loved the southern countryside, the woods, the bramble-skirted, haphazard fields with their scabious and purple-headed knapweed and their linnets, goldfinches and burnet moths. I loved too the little Wealden villages. Villages that had heard the tramp of many feet, the feet of Roman, Saxon and Norman; here, somehow, history and nature seemed to dwell together in a timeless union that to my mind could and should go on for ever.

But now my mind was elsewhere. It was with Random, as she crouched upon her resting place in the sunlit shed. Relaxed, at ease, and growing daily in size, power and beauty, the sheer force of her personality was overwhelming. Even at that early stage in her career (she was less than three months old) she knew exactly what she wanted out of life and made certain that she got it with as little delay as possible. Every morning, when the weather was fine, I would pick her up in my arms and carry her out to her place on the lawn. She was easy to carry. She would lie like a puppy, warm and trusting, her eyes taking note of all around her, and her feet, which still had little gripping power, clinging to any part of my pullover-clad chest that they could reach, like a small child holding its father's hand. She would lie on the lawn, her wings and tail stretched out to the late summer sun in the authentic spreadeagle position, her eyes closed in bliss, the inner eyelid drawn down like a blind, as she revelled in her youth and the sheer, sensual pleasure of being alive. Her tail feathers and the primaries of her wings were lengthening daily and her legs, feathered almost to the toes and rugged as young oak branches, were soon strong enough to support her as she waddled about

22

the garden with her comical, rolling, almost nautical gait.

By now she was beginning to look more like the popular idea of an eagle and less like a cuddly woolly toy from the soft goods counter of a department store. She had, moreover, a pronounced sense of humour, and was as playful as anything of her age should be. She loved sticks and would toss them into the air, trying to catch them in her talons before they hit the ground. She would grab an apple thrown to her and waltz round it, slashing and striking with the speed and precision

of a professional pugilist. She was obviously enjoying her-self thoroughly, although I knew well enough that this play was in fact building up her muscles and developing her co-ordination between foot, eye and wing — a necessity in her future role as a full-time huntress.

She would also spend a considerable time in what I called 'flying on the spot', flapping her great wings with increasing speed until she was lifted a few inches above the ground. Sometimes, as time went on and her wing-power developed, she would find herself floundering about the lawn, just clear of the surface, like a too heavily loaded aircraft. After these exercises she invariably preened herself thoroughly, settling each feather back into place until she was as smooth and immaculate as an eagle on a lectern. I had fitted a pair of jesses to her legs and each day I would take her for a short walk about the village, introducing her to friends and neigh-bours who, quite rightly, admired her enormously and showed a very proper interest in her development. She, for her part, was utterly fearless and at that period of her life was ap-parently ready to be friendly with anyone who treated her with the respect she considered due to her. None the less, her latent power and potential savagery were there all right, lurking just below the surface.

Not long after Random's arrival, when on a quick dash up to London, I acquired a large American horned owl from a pet-shop. The fact that, as is regrettably frequent in such establishments, the price asked for the bird was at least ten times more than it was worth and about twenty times more than I could afford, deterred me not one whit. I am the pet-shop owner's dream customer and can seldom enter one of these emporia without emerging with something expensive, exotic and probably dangerous. This particular owl was young, healthy and very tame, and had obviously been hand-reared.

I have always been fond of owls, particularly of the big, powerful birds with their yellow or orange eyes, which are grouped together under the rather misleading name of eagle

25

owls. They do not look in the least like eagles but more like feathered lynxes or some other ferocious member of the cat family. The American horned owl is a typical member of this group; not as large as the great European eagle owl, it is still an impressive bird indeed and looks even more savage with its brilliant yellow eyes, which in the European bird are orange amber in colour. Its broad, feline head is surmounted by a pair of satanic-looking feathery horns, from which the name is derived, and the whole effect is one of a ruthless wildness, which has to be seen to be believed. Only an adult female goshawk, in full yarak, looks quite such a professional assassin. The appearance does not belie the bird's character, as it is prepared to attack, and frequently overcomes, any small or medium-sized mammal or bird, including other birds of prey, even the redoubtable red-tailed hawk of North America falling an occasional victim to it. The plumage is goshawk grey, with dark bands across the breast, reminiscent of those worn in a similar position by the white-fronted goose.

I called the owl 'Cheetah' because of a fancied facial resemblance to one of those dramatic-looking and speedy predators and installed him in an airy well-raftered loft which had been used for storing apples. Having been hand-reared he was completely tame and after a week or so was allowed full liberty to go where he liked. He used to spend much time sitting bolt upright on a thick branch close to the trunk of what was said to be one of the oldest and biggest mulberry trees in the country. Here, his yellow eyes half-closed and satyr's horns raised inquiringly, he would doze away the daylight hours. Seemingly asleep, he yet missed nothing of what was going on around him. At dusk he would drop out of his tree and float silently, like a buzzard-sized moth, about the garden, seeking whom he might devour.

His favourite observation post was the upright handle of a garden roller and here, grey and silent as a piece of owl-shaped granite, he would wait motionless, listening for the rustle and scurrying of the mice or voles in the rockery, with

scarce a beat of his four-foot wing-span. Before going to bed I would seek him out, give him his nightly meal of meat or liver, and carry him back to his loft, as I feared that he might get into trouble during the long hours of darkness ahead. Cheetah was quite intelligent for an owl, most of which birds, I regret to say, are remarkably slow in the uptake. In fact, the first person who started the legend of the wise old owl was the victim of one of the biggest confidence tricks in natural history. The other popular idea, that owls are dazzled or ill at ease in bright light, is also a misconception, in my experience.

One sunny morning, while Random was lying taking her ease full stretch on the lawn and I was writing letters in the sitting-room, which looked out across the garden to the Weald below, I looked up in time to see a grey shape stooping like a falcon past the window. I dashed out, to find Random and Cheetah locked talon to talon whilst thrashing about all over the lawn. The owl, without thought of the consequences, had gone straight into the attack against an opponent more than four times his own weight.

Luckily for Cheetah, Random's grip had only a fraction of the force it was to acquire with maturity but nevertheless the situation was a serious one. Although outweighed and held at a distance by Random's huge feet and powerful legs, Cheetah had no idea of giving in and was doing his determined best to get his huge adversary into a position where he would have her at his mercy. Random, more surprised than frightened or aggressive and still virtually a nestling, preferred to fight on her back and keep her enemy at a distance. I managed after a long struggle to work both sets of talons apart, but not before Cheetah had managed to send one set of grappling-hooks right through the fleshy part of my left hand. I took extreme care after this episode to make sure that the two rivals had no chance to finish their dispute, which would undoubtedly have ended in the death of one of them.

Random was still apparently a helpless youngster, full of gentleness and good nature. She would feed from my hand and showed no resentment at being picked up unceremoniously and carried to or from her shed. One day, however, I decided to remove the remains of a rabbit on which she had been feeding and which I thought would serve well enough for her next meal. As I reached out for the unlovely remains she suddenly, without the slightest warning, struck out with one foot. She caught my ungloved hand and held on, squeezing convulsively. Her hackles had risen and she uttered a horrible rasping screech.

I was in considerable pain and the blood was oozing out of the holes where her rear and middle talons had gone in. I knew that if I struggled she would grip still harder, and so I just waited, bearing the torture as best I could. After a few seconds she relaxed her hold, her neck feathers subsided, she waddled a few paces away and began preening as if nothing had happened. Resisting the temptation to give her a clip over the ear with a handy plank, I departed rapidly for the bathroom in search of Dettol and bandages. When I returned ten minutes later, wearing a glove forced uncomfortably over the bandages, Random greeted me as if this unpleasant incident had not occurred, clucking gently and tweaking my fingers with her usual jovial good humour. Never again, I decided, would I interfere with Random's meal unless I was gloved to full capacity.

I realized, of course, that I was to blame. The female eagle who feeds her young at the age that Random had then reached always brings the prey and leaves it on a convenient ledge for the eyas to demolish at leisure, departing immediately and making no attempt to interfere or to remove the remains when the youngster has finished or until the prey has ceased to be recognizable as such. This incident was so unexpected and Random's display of jealous rage so disconcerting that I resolved then and there to treat her with far more caution than I had been doing. I had, during the past ten years, reared many young birds of prey, including such close relations to the golden eagle as the tawny and Wahlberg's eagle, but never before had I experienced a display like this.

From that moment on, whenever she was fed, even if she was not particularly hungry, Random would invariably indulge in an unseemly exhibition of bad manners. Grabbing her rations she would shuffle off to the shelter of a bush or an angle in the wall of the house. Here she would mantle furiously, wrapping her wings around the food and giving challenging looks which reminded me of a stage villain in a rather poor melodrama. Throwing back her head and raising her hackles, she would glare at me as if I were her worst enemy, instead of her friend and food-provider. If I were rash enough to approach closer, she would execute what appeared to be a back somersault, turning with the speed of a leopard, and would come shuffling after me, trailing her wings and spreading her tail to its full extent. If I stood my ground and offered her my gloved fist she would seize it with one foot (the other still

29

holding her meal) with her usual chorus of strangled squalls, which for pure ferocity took a lot of beating.

I must confess that at times such as these I began to wonder what the heck I had let myself, and more important still, my family and friends, in for. Ten minutes later, Random having polished off her meal and become once more as gentle as a cooing dove, my qualms would subside. Surely they were merely the signs of growing up? None the less, I could not help thinking of Captain Knight and his two Grampians and wondering if, after all, I would be wiser to deposit Random at the London Zoo and get a nice well-behaved peregrine, or perhaps a goshawk, instead. Then, later, when I saw her standing on the back of a garden seat preening her lovely dark feathers, feathers that had all the bloom of a ripe Victoria plum, and when she turned to welcome me with her ridiculous little falsetto clucks, the undiluted majesty of her appearance made me tingle with joy at the possession of her. To think that this indescribably splendid being, one of the most perfectly designed of all living things, should share her life, if only for a little while, with me and me alone.

Even at this age, gawky and undeveloped as she was, she never lost her sense of dignity. Most young hawks, faced with the unknown, will panic and thrash about at the end of their leashes, damaging their feathers and wrecking their nervous systems. Random, on the other hand, confronted for the first time with a double-decker bus or a carnival procession, would give a hard, calculating stare, her hackles would lift slightly, as would the feathers on her crown, and I could see her making a mental note to deal with the offender at the first opportunity.

The books will tell you that the golden eagle, despite its dramatic appearance and long career as the emblem of empires, is in truth a cowardly bird, sluggish in habit and given to battening on carrion like an overgrown buzzard, only hunting for its living when all else fails. This may possibly be true of the wild eagle, though I doubt it, but it certainly does not apply to Random, who is ever ready to face up to and tackle anything or anyone that could be regarded as a potential quarry. From the beginning of our association she showed complete disregard for her own safety and was prepared to take on all comers, though I must confess that few of her visitors have been prepared to find out whether or not she was bluffing!

Chapter Three

At this period of her life Random spent most of her time either squatting on her haunches with her legs stuck out in front of her or flopping in the grass with her wings spread, her eyes shut, and her head turned at an angle that suggested she was doing her own impersonation of a dead eagle. Indeed on more than one occasion I was half convinced that she had suddenly expired without any apparent cause.

She had the freedom of the garden and could always be found in her own particular favourite part of it. One afternoon, however, returning from a poaching expedition with Tally, I walked through the house and into the garden swinging a recently slain rabbit by the hind legs. Confidently I crossed the lawn to the corner where stood the old green garden seat. Random was nowhere in sight. I looked unbelievingly round, while the temperature seemed to drop a few degrees. There was no sign of her and the garden suddenly felt bigger and emptier for her absence. Before the shock of her loss had really begun to sink in, I heard the strident, churring calls of our resident pair of mistle-thrushes and my hopes rose, knowing well that the aggressively courageous birds were swearing at some intruder in their chosen territory. There was Random, perched fifteen feet up on the roof of the apple loft, which was the diurnal roosting-shelter of Cheetah, the horned owl. She had never shown the slightest inclination to leave the ground before but now she was happily aloft, apparently full of self-satisfaction at her achievement. By some curious optical illusion she appeared even larger than she had when at ground level. When she saw the rabbit she leant forward, her wings half opened, looking more like a lectern eagle than ever, and began to call querulously, apparently expecting me to take wing and land on the roof beside her. When I failed to do so she gave me a look of disdain. It was obvious that her opinion of me had dropped to zero.

Since she either would not or could not return to earth under her own steam and I wished to end this somewhat unsatisfactory state of affairs, I fetched a ladder, which was, of course, just too short for the purpose. However, I managed to struggle onto the precarious-looking guttering. Swaying with vertigo, I reached her somehow, gathered her up in my arms like an enormous hen, and, feeling as triumphant as Hillary on the summit of Everest, returned thankfully to the ground. There Random, her contempt forgotten, was soon tearing happily at her food as if this embarrassing incident had never occurred.

It was obvious that at this stage of her development she could fly upwards but not down again, and I was concerned lest she might land in a tall tree from which it might be impossible to retrieve her. Therefore, I fitted a long leash to her jesses and attached the end of this to a specially chosen branch, selected to give the maximum comfort to her feet, being of varying thickness so that it was as much like a natural perch as possible. I also gave her a big flat-topped log, which I placed at the furthest extent of her leash, so that she could exercise her wings, bounding from one perch to the other and back again.

The time had come when her training must begin in earnest. From the start of this training she proved herself to be both co-operative and remarkably quick in the uptake. In a short time she would jump from the top of a five-barred gate or similar convenient launching-pad on to my left fist, to receive her reward of fresh meat. My main intention at this time was to teach her to fly down to the ground from increasingly elevated perches. Luckily, in a field nearby was an ancient fir tree which might have been grown for this purpose. It was about thirty feet high, with branches which, starting about six feet from the ground, sprouted with almost ladder-like regularity

33

for the next fifteen feet or so. All I had to do was to hoist Random up to whichever level her increasing prowess entitled her.

When she was comfortably settled, had made a leisurely examination of her surroundings, and had begun to call to me to let me know that she was ready, I would walk quietly to the middle of the field, the grass of which was kept short by the grazing of a small flock of southdown sheep and by more than its fair share of rabbits. About a hundred yards away from the foot of the tree I would turn to face Random, calling her by name and throwing down the lure. This lure was the most important piece of equipment that I as a falconer owned. It was the one controlling link between me on the ground and the distant Random, erect and watchful on her perch, close to the summit of the solitary pine tree, which loomed, black and gaunt, against the sky-line. The lure itself was not a particularly attractive object, except, of course, to Random. It consisted of a pair of my mother's superannuated rabbit-skin motoring-gloves, with the fur, or some of it, still attached. These gloves were stuffed with wadding and lashed together until they bore a faint resemblance to some long defunct rodent, hitherto unknown to science. A pair of leather boot-laces fastened juicy pieces of raw meat to this curious apparatus.

Of course, Random had been allowed to feed from this lure for some time before the flying lessons started and she knew what it was for and what to expect when it was produced. She had her own ideas about it, as about everything else. She soon found out that, by holding it down with one foot, she could easily remove the morsels of meat with her beak. Having done so and eaten them with relish, and seeing that no further rewards were forthcoming, she ripped the wretched thing to shreds. When, unwisely, I tried to rescue the grisly remnants, she came bouncing after me and chased me round the garden, much to the merriment of some visitors who happened to be watching from the drawing-room window. This, apparently, was all part of her rather bizarre sense of humour. Certainly she did not expect me to hold it against her. Her moods changed like the drifting shadows on the lawn that was her daytime playground. This temperamental nature ensured that, whatever else it might be, our relationship was never dull.

Random, poised upon her lofty look-out and standing on the tips of her toes, would watch my every movement as I stood below her, swinging the lure and shouting. Her head would bob, her wings, half unfurled, would feel for any trace of breeze that would help to bear her up, and finally, excitement and greed overcoming her uncertainty, she would drop out of the tree, dark and elemental as a storm cloud, with two powerful downbeats of her immense wings, followed by a long dramatic downward glide, which would bring her close to where I stood. There she would fan her tail to its fullest extent and drive her wings forward until the tips seemed almost to meet beneath her body. Her great yellow feet would shoot out in front of her as, with a sudden rolling plunge, she would drop upon the lure, grab it with one foot, glare defiance at me, and, finally, convinced that she had instilled a proper sense of respect, take her pleasure of the meat with which it was garnished. I would wait until she had gulped the last morsel, roused her feathers and was looking for more food, before 'making in' quietly to pick her up and attach her leash to her jesses.

Sometimes, in a ritualistic display of aggression, she would drop the lure at my approach and spring several feet to my thickly-gauntleted fist, which she would then squeeze convulsively until my whole arm became numb with the force of her grip. On these occasions I would give her a specially delicious morsel which she would gulp down contentedly while still gripping my fist with all her power. She would soon relax her stranglehold on the glove, her raised hackles would subside, and I would stroke her gently on the breast while she twittered to me in her falsetto baby voice to show that she really was still my friend and that the previous display of bad manners had been a bit of ham acting. None the less, I was thankful that my hand and forearm were protected by two thicknesses of impenetrable hide.

After these training sessions I would carry her happily homeward and release her in the garden once more. Of course, these flying exercises, which were intended primarily to build up her muscles, increase her confidence and aerial dexterity and above all develop the bond between us, could only be carried out when she was hungry. Random is basically bone idle. Her attitude has always been that – whilst sailing about the sky on thermals, looking noble and authentically aquiline, may be all right for other eagles – she would rather sit looking noble on a nice, comfortable branch with one foot tucked into her breast feathers and a pleasantly full crop, dreamily contemplating the beauties of nature below her.

I, however, had different ideas. I was determined to make her do something to justify her existence and the enormous meat bills I was running up. We eventually arrived at a satisfactory compromise. All she had to do to earn her supper was to fly; not, one would think, much to ask of an alleged monarch of the air. She had been accustomed to receiving her main meal at about three o'clock in the afternoon; she was more than ready for it by then, as, without being actually gluttonous, she was none the less a remarkably 'good doer' and her feeding time was the high spot of her day. Therefore, at three-fifteen or three-thirty, when she was feeling thoroughly 'sharp-set' and ready for anything, I would continue her education. Soon, as her wing-power and self-confidence grew, she would come many hundreds of yards with the minimum of hesitation.

To see her launch out from the top of some heavily foliaged tree, bursting like a shell through the greenery, her great broad wings smiting the air and her head thrust forward like the prow of a viking ship as she gained impetus for her final plunging stoop to earth; to see her growing in stature as she ate up the distance between us; was worth all the care that had gone into her early training and the worry as to how she would react when she finally found herself free of restraint. Sometimes I would hide the lure just before she arrived to clutch it and she would wheel round me in a half circle only a few feet above my head. Her eyes would look straight into mine as she passed and the long primary feathers of her wings, stark against the sky like probing fingers, would stroke the air as she banked and swung past before crashing into the depths of another tree close by. When the lure was finally produced she would heave herself into the air and fall upon it after one long gliding stoop which carried her straight to her target.

Random had now reached her full stature although of course her juvenile plumage would remain with her, being renewed during her annual moult but each time becoming more like her future, mature garb, until her fifth year when she would be fully adult. She now measured an inch or two under three feet from beak to tail tip, her wing span was fully seven and a half feet, and she weighed about twelve pounds, being therefore some four pounds heavier than the male of her species,

which is noticeably smaller all over. Her feet were armed with immensely powerful black talons, the hinder ones on each foot being almost frightening in their lethal strength; her back was broad and springy to the touch; and she had a chest of which a bull-terrier need not have been ashamed. Her eyes, which with maturity would become a clear amber colour, were now deep smoky grey and she gazed about her with all the fearless arrogance of one who knows full well her own potentialities. So formidable did she appear that her many visitors regarded her with an awe with which I, having seen her in all her moods, could fully sympathize. Strangers meeting her for the first time greeted her with the sort of nervous respect that many people reserve for alsatians – only more so; after all, she could fly.

Watching her at one of her favourite games, demolishing a large cardboard box, which she rapidly reduced to postage-stamp-sized fragments, I couldn't help wondering what would have happened had it been Random rather than Cressida, the kestrel, that I had found on that beach at Camber in 1942. What would have happened had she accompanied me to the battlefield at Sejanane? What would have been the reaction of my captors and later custodians?*

About this time I decided that it was high time Random was fitted out with a new hood and, more important still, taught to wear it with dignity and aplomb. The hood, like the lure, is a piece of equipment without which no self-respecting hawk, falcon or eagle would consider her wardrobe complete. The hood itself is a skilfully fashioned and beautifully decorated leather cap which slips over a hawk's head, plunging her into utter darkness; thus, she can be carried about under conditions that would probably fray her nerves were she able to see the horrors of modern civilization all around her. Being hooded, she feels secure in her little private world of artificial night and can be carried undismayed through city traffic, can travel by train or motorcar and will arrive fresh and eager for

* The story told in *Lure of the Falcon* (Collins, 1972).

38

anything that may be afoot. The hooded falcon, carried to or from the hunting field, does not waste her energy by bating at any prey, however unsuitable, that may get up within range and is thus ready, eager and unfrustrated, when the time comes to cast her off from the fist to deal with her own legitimate quarry. It is also useful to be able to hood her up when she has to have her beak or talons coped, her jesses renewed, or other indignities perpetrated upon her person. Incidentally, the hood is used almost invariably with falcons, less frequently with goshawks, and only very occasionally with sparrow-hawks and other small 'short wings'.

My only previous experience of hoods and hooding had been unhappy. A friend of mine in Kenya had acquired an eyas Ayer's hawk eagle which he trained efficiently. However he decided to 'make her to the hood'. Hoods being unobtainable in that part of East Africa at the time and my friend being no craftsman, he wrote to me in England asking if I could get him one and send it out to him. I wrote to the secretary of the British Falconers Club who had one made for me, fitted to the measurements I sent him. It was a beautiful thing, a real work of art, and I was sorely tempted to keep it myself, if only as an ornament for the mantelpiece. However, friendship prevailed, and I decided to send it on.

I was not entirely happy about the fit, and there being no handy hawk eagles available to try it out on and the London Zoo being none too enthusiastic about experimenting on their only specimen, I managed to persuade the Curator of Birds at a certain well-known museum to let me have a try on a mounted example of the species. The Curator stressed that the specimen was rare, valuable and possibly irreplaceable, and counselled me to use great dexterity and gentleness of touch. Having said this he skilfully eased off the glass case in which the rather sorry-looking bird was housed and bid me do my stuff and be quick about it. Hooding was no problem, the specimen in question offering but little resistance. The hood slipped on as if designed for that very purpose, as indeed it had been. Delightedly I gave a quick adjustment to the braces which tightened or loosened the hood when on the bird's head. Without thought, I gave an admiring tweak to the handsome plume of golden pheasant feathers that adorned the top of the hood – and head, hood and all came off in my hand. The gentle Curator's oaths were terrible to hear, although he eventually calmed down and was very nice about it. This was a long time ago and I have not paid a visit to that particular museum since.

I took the measurements of Random's hood and presented myself at the workshop of a local cobbler, who confessed that he had never seen a live eagle, let alone an eagle's hood, but, good sportsman that he was, was ready to do his best. I lent him an ordinary, rather time-worn peregrine's hood of the Dutch variety as a model and he told me that he would get to work on it at once, during the evenings, when his normal daily work was done. It took a long time but the result was well worth waiting for. The hood, when I eventually received it, was an example of an English village craftsman's art at its best. Shaped much like the helmet of a medieval knight, with an opening for Random's beak in lieu of a visor, fashioned out of rich, mahogany-coloured leather, with two eye-pieces of deepest blue, and surmounted by a cockade of feathers from an Old English game bantam, it was indeed a brave sight. The hood fitted comfortably into my cupped hand and was amazingly light but firm and crisp and obviously capable of standing up to a good deal of rough usage in the field.

All I had to do was to persuade Random to accept it as part of her daily round. This was a task to which I did not look forward with much glee. I knew too well how she could behave when in one of her tantrums. She was sitting in one of the lower branches of a small sycamore tree when I sought her out. Catching sight of the hood, she turned her head on one side and gave it a quizzical look, apparently uncertain as to what its function was supposed to be, but hoping, perhaps, that it was some sort of new-fangled lure. She appeared to be waiting for me to throw it down for her inspection. Wishing to get over as soon as possible what I was sure would be a particularly painful confrontation, I picked her up, carried her to a sheltered part of the garden, and with my right hand, in one movement, quietly slipped her beak through the opening of the hood which I then quickly eased back until it fitted snugly on to her head, tightening the braces with one hand and my teeth. That done, I tensed myself for the display of fireworks which I was certain would follow.

To my amazed relief no pyrotechnics were forthcoming. Her grip tightened alarmingly on my wrist. She bent forward gropingly to bite my glove, perhaps to make sure that it was still there. She shook her head violently until one of the braces caught my left eye with a stinging blow, which made it water painfully for a moment. She scratched the eye-pieces thoughtfully with each curved rear talon in turn, just to make sure that her head was still attached to her body. Then she gave up. She had tightened all her feathers around her, so I knew that she was still feeling far from at ease. Wishing to reassure her, I stroked her breast with my right hand, and her foot flashed out, striking forwards like the forefoot of a vicious stallion. Luckily my reactions were quick and she missed her mark, but I decided to put her down on her perch to rest and to become accustomed to her new burden.

She looked magnificent with the sun shining on her lustrous, plum-glossed plumage, the feathers of her hood fluttering in the breeze. I went indoors to call my mother and some friends who were staying with us so that they might share this picture of medieval splendour. On our reappearance, less than five minutes later, Random was sitting where I had left her, one foot tucked up as she preened contentedly. The hood was lying on the grass below, where she had thrown it contemptuously aside. How she had removed it, I knew not, but removed it she had, and in a remarkably short space of time. I put it on once more and she bore the hooding as calmly as before. She might have been made to the hood for years, so philosophically did she accept this new situation.

41

Chapter Four

Tally arose with a yawn and a long luxurious stretch, which rippled up the length of his lithe, muscular body. He had been lying in the sun with fore-paws gracefully crossed, as was his wont, and now he came forward in the hope of some action. I looked at the pair of them, each perfectly designed to fulfil their purpose in life, to hunt down and capture their natural quarry, the one in the air, the other speeding on foot across the scrubby, ravine-traversed deserts of the Middle East. Gazing at them as they stood close together, Random's hood pennant fluttering in the breeze which stirred the silky feathering on Tally's proudly lifted tail and long, delicate ears, it struck me that they might have stepped down from some medieval tapestry depicting the departure of a hunting party. I decided there and then that I would train the pair of them to work together, so that I could enjoy watching them doing that for which, presumably, they had been put on this earth, and doing it in partnership.

For hundreds if not thousands of years the saluki has been trained to co-operate with the saker — that swift, beautiful desert falcon — in hunting the Dorcas gazelle across the sands of Saudi Arabia. Before our arrival in England, Tally and Yimkin (my passage* male saker or sakeret) had shown a natural affinity which was fascinating to watch. Yimkin had been given to me by a young farmer in Kenya who had caught him on his farm near Kitale in the north of the colony. The farmer had been interested in falconry but, not having the

* This term applies to a wild-caught falcon in its first year
before assuming adult plumage.

42

time to practise it properly, was sensible enough not to try to keep a bird to which he knew he could not do full justice. He had seen Karen, my Eleanora's falcon, in action, and knew that I would be able to give this delightful bird the attention and exercise he needed. I used to fly him 'at bolt' or straight from the fist to the quarry (chiefly francolin) after the manner of a goshawk without the preliminaries of waiting for him to gain height, so as to deliver the death-dealing stoop that is the most dramatic sight in the whole realm of falconry.

Flying such a falcon 'at bolt' may not seem particularly sporting or exciting but I was stationed at the time in a game reserve where shooting was prohibited yet where nothing was said about falconry, because, I suspect, the authorities had no idea that such a pursuit still existed. I liked to eat, was disenchanted with the products of Messrs. Heinz, and thus Tally and Yimkin were the best and most natural means of keeping me (and them) supplied with fresh game; besides, a youngish francolin, properly cooked, is about as delicious an addition to the menu as one could wish for.

Without any deliberate training or previous experience, Tally would watch Yimkin's line of flight and streak after him, keeping him in sight until he either 'bound to' and brought down his quarry or else missed it, gave up the chase and took stand in an acacia tree. If he came up when Yimkin had taken a quarry and was about to break into it, he would leap about barking loudly and wagging his tail in excitement, although he was not normally a noisy dog. This rather disconcerting habit of barking under these circumstances did not apparently upset Yimkin and proved exceedingly useful in telling me where the next meal was awaiting me.

There was a strange if superficial similarity between Tally and Random. Both were big, handsome and powerful. Both gave an immediate impression of surging speed and strength with just a hint of primitive savagery, controlled, but only just. Each was an obvious aristocrat with an air of timeless good breeding and each was somewhat aloof and inclined to be standoffish with strangers. They more or less ignored each other, possibly realizing that in a trial of strength it would be, as the Duke of Wellington said about the battle of Waterloo, a damned near-run thing. For my part, wishing to build up an affinity between the two of them and with the idea of encouraging them to work together as a team, I did try my best to make sure that nothing occurred that might spark off any latent hostility lurking beneath the surface and precipitate a conflict.

None the less, the conflict took place, as I suppose it was bound to do. It was, of course, my fault. Random had eaten her fill of the hind-quarters of a hare, a roadside casualty which, unfortunate as it was for the hare, was a welcome piece of jetsam. Such a hare was enough to provide the eagle with meals for days and also supplied the necessary 'casting' (fur and bone) which is essential to all birds of prey – keeping the digestive organs in full working order and removing any grease and slime from the crop, finally to be ejected from the mouth in the form of a firm, hairy, egg-shaped pellet. Such pellets could always be found under Random's roosting-place and were one indication that she was in her usual good health.

I had left the meaty, russet-furred remnant on the lawn when I had to run in to answer the telephone, and had not bothered to find out exactly where Tally was lying. I came out a few minutes later just in time to see him trotting across the lawn to examine what he must have thought was something put there especially for his delight. As he picked it up in his mouth and was moving away towards the shrubbery, probably intent on devouring the savoury morsel at leisure, Random came to sudden life, leapt from her resting-place on the back of the garden seat, and covered the few yards between them like an avenging angel. Her armed feet struck Tally amidships and bowled him over as if he had been shot. Tally weighed about seventy pounds to her twelve and a half, but taken by surprise he didn't stand a chance. In one movement, so swift and certain that it might have been the result of long practice, one of Random's feet grabbed Tally's muzzle in a grip of iron, completely immobilizing it, avoiding as if by instinct his slashing riposte. The other foot took hold behind the shoulders and she had him at her mercy as she tried desperately to bend his body into a bow — and so by pure muscular power of beak and foot to break the vertebrae and finish the fight. Tally, thrown off balance and with his jaws out of action, tried to bring his hind legs into play and thus to disembowel his opponent, but she was holding him from behind and he couldn't get sufficient purchase on the soft ground to roll over and crush his opponent beneath him.

All this took seconds, the time it took me to cross the lawn and, without thought, to do the best thing that could have been done. I grabbed both Random's thrashing wings and she, finding herself, as she thought, attacked from the rear, let go her two-footed grip and rounded on me as Tally picked himself up and fled to the house. When Random released her grip on her opponent and turned to deal with me, I offered her my glove, which she seized at once, swinging herself on to it with an annihilating clutch as I gripped her jesses with my other hand. I hoisted her up and somehow settled her, still clutching spasmodically, on to my fist, where after a moment or two she calmed down. I was able to fasten her leash to the jesses, putting her back on her bow perch where I left her to her own thoughts.

Tally was lying in the hall, licking his back from which the blood was oozing in deep red blobs. By the mercy of providence his wounds were superficial. His jaws were bruised and stiff but the great foot that had held them had completely encircled them, acting as a muzzle without penetrating the sensitive flesh at any point. There were two nasty-looking wounds where the two central and hind talons had gone home in his withers, but here again luck had held. Tally's thick shaggy coat, tough rubbery skin and muscular development had saved him from anything that a strong antiseptic dressing could not put right. Had Random reached her full strength it

45

might have been a different story. It is a fact that even the male golden eagle has been known to kill foxes – and the Siberian sub-species, bigger but not all that much bigger than Random, is regularly used in central Asia and Turkestan to hunt wolves. One of these birds is known to have accounted single-handed for thirty-six wolves before losing a foot when tackling the thirty-seventh. This eagle was retired from active service and later received a state pension, proving that the Soviet authorities can be human, after all.

It was now painfully obvious that my dream of watching Random and Tally hunting together could never be realized. On reflection, however, this was probably just as well. Such a combination would hardly have been a sporting proposition; no hare, however swift and artful in evasion, could have survived such an overwhelming force. An experienced saluki, even operating alone, can out-run a full-grown hare. A golden eagle, too, once it has gained its full powers, has little difficulty in overcoming this wily quarry, anticipating its every movement and flying it down by sheer speed and determination, despite all the jinking and turning that the fugitive can produce.

Apart from the sudden frightening fits of anger (always triggered off by some sort of jealousy, either gastronomic or territorial) Random was surprisingly placid and seldom showed resentment at being admired or even stroked by complete strangers. In fact, she showed so much common sense and seemed so thoroughly at home in her own surroundings that I decided that she should remain 'at hack' for as long as possible (which meant, until she showed an inclination to wander off or to attack something or somebody that she might consider an intruder).

'Flying at hack,' briefly, implies allowing a hawk or falcon to remain at liberty, taking its own exercise and following its own whims whilst returning regularly to a certain area for food, until it shows increasing independence and puts in less regular appearances at its feeding station, showing that it is now catching its own prey. Such a period might last for a few days up to several weeks. Once it is over the time has come to start the hawk's serious education. Eyas peregrines, merlins, and even sparrow-hawks are all suitable subjects for this form of temporary liberty, being small and elusive enough to avoid attracting an unnecessary amount of attention to themselves from possibly hostile observers, but with a bird as large and powerful as a golden eagle I feared I might be letting myself in for rather more than I had bargained.

However, there was one great advantage – Random was already almost fully trained. She was so tame or 'well-manned' that she could be taken up at any time and never showed the slightest inclination to stray away from her home. Fifteen years later she is exactly the same and can be released to amuse herself and any admiring spectator for hours on end, in the knowledge that when I wish I can call her down from wherever she happens to be sitting. The only difference now is that, as she is a fully mature adult with considerable hunting exper-ience, one has to ensure that all small domestic animals and birds (such as dogs, cats, poultry and wives) are safely enclosed before she launches forth to enjoy herself in her own spectac-ular way.

She soon learned that the car was the open sesame to the excitements that lay beyond her own private domain. She would sit placidly on the wooden block provided for her on the back seat; as we sped along country lanes or through villages she would peer out at the world flashing past the windows, craning her neck and calling loudly if something of particular interest caught her attention. The sight of an enormous eagle sitting happily in the back of the car caused amazement and some consternation amongst the inhabitants of the few scattered villages through which we passed. It used to amuse me immensely to see the startled gaze of drivers of overtaking vehicles when they suddenly realized that what they had been idly contemplating and which had probably appeared to them as a curious eagle-shaped object, perhaps some piece of antique statuary, was indeed an honest-to-goodness eagle and very much alive.

We used to take her to the wide, low-lying, rushy fields that bordered on Romney Marsh, and let her wing her way to a tree where she would stand erect as on sentry-duty while I explored the country below, watching the shovellers, the dabchicks and the rare fifteen-spine stickle-backs that lurked in the brackish reed-fringed waters of the dykes. Random would wait, following my every movement from her lofty station, sometimes calling out to me in her high-pitched, yelping voice. She would often drop out of her tree and sail around just over my head, watching the ground below in case something should arise that needed her immediate attention. After one of these great, sweeping, expeditionary flights, she would pitch in another tree or perhaps land delicately, with half-open wings, on the frail summit of a low hawthorn bush, where, as the springy spires bent beneath her weight, she would sway with fanned tail fully extended, finally letting go to land with a swishing thump beside me.

These forays into the wide, flat country far to the south of our home were great fun. We had the marshes to ourselves with no prying eyes or strident voices to spoil our enjoyment of each other's company. Sometimes I would take Tally as well as Random. I had worked out a method of keeping them well separated on the journey to the chosen venue. In any case, after that one brief but painful unpleasantness, they now appeared to regard each other with a watchful, if somewhat uneasy, neutrality. Tally had no intention of being caught off guard again and Random knew this well enough.

Arrived at our rendezvous, I would first release Tally for a good hard gallop, which he, being young and full of life, loved and needed more than anything in the world. He would set off at his tireless, rhythmical, half-extended gallop, spring-ing over the tussocks, clearing the dykes, his eyes ever alert for the russet-red and snowy-white form of a scurrying hare. The marshland was dotted with sheep, of the spongy-hooved, thickly-fleeced, Romney Marsh breed, the only sheep that could stand up to the semi-waterlogged conditions that were their daily portion. By merciful providence and for some reason known only to himself, Tally did not appear to regard these sheep as being worthy of his attentions – in fact, he hardly seemed to notice their existence. This might well be because the sheep seldom ran from him. They would gather in tight formation and merely stare at him as he raced round their flanks, nose aquiver and eyes alert for more sporting quarry. Had his attitude been otherwise, these expeditions would have been virtually impossible. After ranging far and wide, travelling almost invariably in huge circles with myself the pivot, during the course of which he must have travelled miles, he would return at last, leering happily, tongue lolling, his coat covered in burrs and grass-seeds. After he had taken a long drink from the nearest dyke, I would open the car door and he would leap in to flop down on the seat, panting and throbbing like a dynamo. I would then release Random, who had been watching Tally from afar and waiting her turn with little patience. Once she was airborne I would busy myself removing the assorted vegetable matter from Tally's shaggy black coat and long pendulous ears.

One afternoon, Tally having lamed himself by running a bayonet-sharp thorn into his tender pad, I took Random out alone. Without realizing it I strayed on to the property of a surly and suspicious sheep-farmer. Random had a long-drawn-out but unsuccessful flight at a three-quarter-grown leveret, which, after an exciting chase during which Random had practically worn herself out, had shown her a clean scut, disappearing into the recesses of a tangled thicket of hawthorn, elderberry and old man's beard – a thicket which he must have used as a refuge many times before. Random was sitting sulking, hidden in the crown of a stunted, wind-distorted but thickly foliaged tree, whilst I, my legs below the knee thick with green slime from a dyke into which I had inadvertently plunged while trying to keep the chase in view, was pleading, cursing, cajoling and waving the lure about, trying to per-suade the contemptuous eagle to leave her leafy fastness and come down, so that we could go home.

I was thinking longingly of crumpets oozing with butter and lashings of boiling hot tea when I suddenly realized that Random and I were no longer alone. Two men stood eyeing me in a distinctly unwelcoming way. One, the obvious em-ployer, was, equally obviously, a typical townsdweller posing unsuccessfully as a countryman. He asked me briefly and rudely what the hell I thought I was doing on his land. I replied apologetically that I was trying to lure an eagle down from the tree under which we were standing.

'Eagles,' he exploded, 'there are no bloody eagles around here! Now, get off my land!'

He looked at me and at the dirty, dishevelled lure in my hand. He was obviously wondering just how long I had absented myself from the local asylum, when Random, who was in a filthy temper to begin with and obviously disliked the fellow's looks as much as I did, burst like a tornado out of the tree and came straight at him. She was indeed a terrifying sight, even to me. How she appeared to the enemy I can only guess. I could see that she meant business. There was quite a different expression in her eyes, which glowed with a look of hate usually seen in those of a charging leopard.

Instinctively, and unwisely, I threw myself between her and her target, who was standing apparently immobile with horror as this wide-winged thunderbolt fell upon him. Seeing that I was about to receive the full weight of her onslaught, she tried to brake, her wings came forward and her tail fanned wide to slow her up, but she had been travelling flat out and she caught me waist-high and slid down to the ground, her talons raking a long strip from the old parachutist's blouse I was wearing. As she hit the ground she tried to dodge round me to tackle the still apparently paralysed foe on foot, but I managed to grab her jesses and with one desperate movement swung her up to my fist, where I attached the leash and slipped home her hood.

The two erstwhile truculent onlookers came to some sort of life and withdrew silently and rapidly in the direction from which they had come. I hurried Random to the car, anxious to put as much distance as possible between ourselves and the scene of this unpleasant and potentially dangerous happening. Odd as it may seem, I heard no more of the incident. Possibly the two men either did not believe or did not wish to admit that it had ever taken place.

Chapter Five

Random's fame and popularity were now spreading. I was frequently receiving deputations of interested local people. Schoolchildren, members of the Women's Institute, policemen and secretaries of natural history societies all presented themselves at our door, and it was not long before we were asked to accept our first public engagement. This was to be an informal talk and flying demonstration at a local public school. I had done a certain amount of this sort of thing with Cressida but none the less I looked forward to the event with little enthusiasm. However, when the time came Random, as I might have guessed, carried the hour with dignity and good sense; behaving impeccably, she did all that could have been expected of her.

I was getting used to seeing her about by now and having watched her grow from a scruffy and, it must be admitted, rather out-at-elbows eaglet into a beautifully proportioned bird, with her subtle colouring and grace of movement, and with a personality to match, I suppose I was beginning to become blasé, to take her for granted. I was therefore always highly gratified to see the reactions of those who were confronted with her for the first time. Random, for her part, always appeared, consciously or otherwise, to play to the gallery, striking the authentic 'queen of the air' attitudes, glowering regally and showing her splendid rapacious-looking profile at the slightest opportunity as she stared with studied indifference into the middle distance. She would spread her seven and a half foot wing-span as she half leaped, half flew, from the back of a chair to my extended arm or performed her special trick of flying up from the ground to catch the thrown lure, turning on her back with amazing dexterity to grab it with a swift upper-cut before dropping to the ground once more, where she would mantle her inanimate prey with an awe-inspiring display of simulated ferocity.

She soon perfected a much more spectacular version of this party piece. I would call her from the upper branches of a tree and when she was coming down fast with her long, slanting semi-stoop, I would throw up a piece of meat weighing about two ounces as high and as straight as possible. She would literally stand on her tail like a Spitfire and climb vertically twenty or thirty feet, shoot out one foot and snatch the morsel from the air. Then she would straighten out and, with the meat still clutched beneath her tail, swing into another tree some distance behind me, where, after swallowing the meat in a couple of gulps, she would turn round and prepare to repeat the performance. She would do this several times until I suspected that her enthusiasm was waning. Then I would throw out the lure, she would do a series of corkscrew turns and fall upon the expected reward, which she covered from view behind the palisade of her drooping tent-like wings as she tore into the juicy goodness which she so well deserved.

Random was now about eight months old and to my delight had settled down into a comfortable daily routine, which seemed to suit the pair of us admirably. She still slept at night in her outhouse and was released early in the morning to take her pleasure in her own leisurely fashion about what she had come to look upon as her own private world. In the afternoon, if the weather was reasonable (because she hated and refused to fly in rain or in misty weather), I would take her up and carry her to her training ground. There were several of these, all topographically slightly different, as I wanted her to become accustomed to flying to me with confidence and without hesitation at any time or anywhere, whether we were alone together or if, as often happened, there was a small excited crowd of eagle-watchers present.

No one in that part of the world had ever before seen an eagle close to apart, perhaps, from a moth-eaten, broken-feathered travesty in some zoo or menagerie, and the chance to see Random enjoying her ease, to take photographs and perhaps even to stroke her, resulted, in the first few months of our partnership, in a steady stream of visitors beating out a path to our front door. Random would allow her breast or crown to be stroked gently and would even sometimes absent-mindedly nibble the fingers of the delighted onlookers, who were amazed at the gentleness and forbearance of what they had expected to be the embodiment of tameless savagery.

Without warning all this changed. One day, a visitor, a lady who had met and, so I believed, come to some sort of under-standing with Random, arrived and asked if she could take a few photos of her. Random posed with her usual dignity and restraint and everything seemed exactly the same as usual. As a gesture of friendliness and appreciation the lady reached out to stroke her under the chin. I noticed Random's hackles lift slightly and a curious, speculative look come into her eyes, but thought little of it. Our guest hadn't noticed these signs and gave her an affectionate parting touch, allowing her fingers to rest momentarily on Random's broad back. Without warning, Random spun round and lashed out with the speed of a striking cobra. Luckily the visitor's reactions were quick and she pulled back her hand just in time, but one talon, none the less, caught her sleeve a glancing blow. The visitor was obviously shaken and so was I, but, as she was wearing a fur coat at the time, I assumed that the eagle's sudden hostility must have had something to do with that. I made my apologies and the lady, her composure regained, picked up her camera and withdrew.

The next morning, two boys from Sutton Valence school called on us. Random was sitting on her block in the sun and, luckily, was leashed to her perch. When the boys were a few yards away Random, who had been watching their approach, motionless except for her head, which she slowly lowered like a bull about to charge, suddenly, without a hint of warning, flew straight at them, pulling the heavy log behind her in her determination to make contact with what she obviously regarded as unnecessary disturbers of her peace. From that day onwards, with but few exceptions, Random has never allowed anyone other than myself to handle or even approach her when on what she considers to be her home territory. She has decided that she is, and will remain, a one-man eagle and will have no truck with anyone else.

The fact that she has known someone for years makes no difference. She is utterly relentless and single-minded in this somewhat embarrassing peculiarity, and no one, by feeding or familiarity, can win her affection, no matter how hard they may try. Whilst this is admittedly flattering for me, her chosen partner, it can be, and often is, highly inconvenient, especially if I have to leave home for a few days and Random is not to be included in my entourage. It is virtually impossible to find anyone, even one's nearest and dearest, who is prepared to risk total annihilation or even the possibility of a few weeks in hospital. Admittedly, confidence can be restored to some extent to the potential eagle-sitter by ensuring that they are suitably clad against the worst that could happen, but unfortunately suits of armour, even second-hand ones, are hard to come by these days.

The result of all this is that if I have to be away for more than a day or so Random comes with me. Otherwise I stay at home. Luckily for me and for Random, the number of people who are prepared to put up with or even to welcome an eagle as temporary lodger in the bathroom or summerhouse is surprising. Away from home she is sweetness itself, provided no one tries to take liberties with her, and she can be released in public without risk of her attacking anyone (at least, up to the time of writing we have not lost a single spectator).

Chapter Six

One evening, casually looking through the small ads in a livestock paper, I saw something that for no particular reason caught and held my attention. The advertisement said something to the effect that a thriving pet-shop, complete with living accommodation, was being offered on lease. The shop was said to be on a busy thoroughfare, close to an attractive park. For further details, one was invited to write to the box number which was included. With little thought for the consequences, I groped for pen and paper and had written for further particulars within minutes of seeing the advertisement.

Having slipped the letter between the sardonically grinning jaws of the pillar box at the corner of our road, I immediately wished I had not been quite so impetuous. What the heck would I do in the East End of London? And, a thousand times more important, what would Tally and Random do there? However, I consoled myself with the thought that merely writing could do no harm. After all, I did not have to buy the confounded place. Besides, seeing the posh address on the Basildon Bond notepaper I had used, the potential vendors probably wouldn't even answer.

However, within forty-eight hours of posting the letter the telephone rang and a gravelly, cockney voice was telling me all about the delights of Bermondsey – for that, it appeared, was where the emporium was situated. The fellow burbled on about the trade figures, which, if they could be believed, sounded satisfactory enough – in fact, so satisfactory that I made so bold as to ask why he was anxious to part with this crock of gold. He told me that he and his brother, who ran the shop, wanted to give up the livestock business, having made what appeared to be a successful killing, and to take on a pawn-broker's business nearer their home in Blackheath. This sounded reasonable enough. The rent of the premises seemed none too exorbitant and all that remained was for me to see the place and to come to an agreement regarding the value of the goodwill and the stock. If our meeting came to a successful conclusion, I would be able to move in and take over in no time at all.

My mother and everyone else I spoke to about my prospective entry into the world of commerce thought that I was quite mad and had no hesitation in saying so. What did I know about trading and in the East End of London, too—a maelstrom of vice, violence, corruption and gangsterdom, or so they all assured me. I explained that, whilst I knew nothing whatsoever about trading, I had at least a modicum of knowledge about animals. I also pointed out that anyone who is prepared to spend money on buying pets for themselves or their families was unlikely to be a card-carrying member of the local branch of the Mafia.

The following morning, very early, I took Tally for a walk in the fields near our house and, making certain that Random was safe and comfortable on her block in the garden, I set out on my journey into the interior. The shop was in one of the many colourful streets just inland from the south bank of the Thames and close to the heart of dockland. I left the bus a few stops too early, as I had decided to walk the last part of the journey, so that I might catch something of the atmosphere, the 'feel', of what might be my home for some time to come. I was pleasantly surprised as I wandered down the road, looking about at what appeared to me, as I saw it for the first time, almost a new world.

Nothing really conformed to what my preconceived idea of what Bermondsey—almost as much a stronghold of cockneydom as Shoreditch or Hackney—would be like. The houses and shops were seldom more than two or perhaps three storeys high — thus the sky was never shut out and there was none of the claustrophobic feeling which I get amongst the jumble of over-crowded noisy streets with their, to me, rather sinister under-tones, which sneak about around the Charing Cross Road and, perhaps even more noticeably, in Notting Hill Gate. There was certainly nothing sinister or devious about the road. The pavements on both sides were crowded with stalls and barrows, giving a rather jolly, slightly continental atmosphere to the place. Their owners regaled the passers-by with their own brand of spontaneous witticisms. One fellow, for instance, was waving a bunch of carnations in the air, which, according to him, were good enough to bloody eat. The whole lot of them looked like refugees from a crowd scene in *My Fair Lady*.

Eventually I came across a small knot of people gathered round a long table beneath a gaily-coloured striped awning. Sitting on the table, holding a cup of tea in one hand and a banana in the other, sat one of the most engaging beasts I had yet to meet. It was a Humboldt's woolly monkey. Covered in silvery grey fur, the texture of plush, except for his face which was coal-black and bore the wistful, contemplative look of a miniature gorilla; with a prehensile tail considerably longer than his body; and with tiny, warm, black-palmed hands, as acquisitive and yet confiding as those of a small child; he was almost exactly like an animated toy. This was obviously the shop I had come to see and it had as good as changed hands there and then. I defy anybody with the slightest feeling for animals to have felt otherwise.

All this I noticed before, with only the briefest glance at the toucan hunched uncomfortably in a too-small parrot-cage on an adjoining table, I shouldered my way through the crowd and went forward heedless as a ram to the slaughter. The proprietors hurried forward from where they had been lurking in the dusky interior. Instinct must have told them that it was I who had written that first letter and they, knowing a sucker when they saw one, were determined (and who can blame them?) to deliver the *coup-de-grâce* as quickly and as painlessly as possible.

The elder of the two brothers, who appeared to me to bear a strong family resemblance to both Fagin and Sikes, hurried me to his den at the back of the shop whilst the assistant, a lively septuagenarian by the name of Charlie (the Artful Dodger?), slipped across the road with surprising alacrity to fetch cups of tea from the café opposite. He then began his spiel. The shop, he told me, was a gold-mine, a veritable El Dorado; he even showed me the contents of the till, to prove it.

It couldn't have taken him more than five minutes to realize that I was a veritable child in all matters commercial. He could hardly have caught me at a better moment. I had for some time decided to do something constructive. I had been home from Africa long enough and was ready for some new experiences, and this seemed at least an original idea. After all, I could always get rid of the shop if it did not come up to expectations. Business seemed to be fairly brisk — at least, there were quite a number of people walking around, though I must admit I didn't see anyone actually buying anything. It occurred to me that, if the worst came to the worst, there was always Sambo as the Humboldt's monkey was called, and I didn't suppose that the price of a second-hand barrel-organ would be exorbitant.

Over tea and bacon sandwiches (the first time I had sampled this excellent mainstay of the east-end street vendor), we talked business. At least, he talked and I nodded at appropriate intervals and tried to look like a potential Rothschild. The offshoot was that I decided to take over the lease and pay for the goodwill, for a sum which I have no intention of disclosing, as even at this distance in time I blush at the thought of it! However, I had saved a certain amount of money from my East African odyssey and my mother had promised to help out if I really decided to go ahead with this improbable proposition.

The shop itself was long and narrow, the walls lined with shelves on which stood cages of hamsters, budgerigars, guinea-pigs and other rather uninteresting beasts of the type usually found in such establishments. The front of the shop consisted of wooden shutters which had to be pushed up and down with a contraption that looked rather like a fishing gaff; this was also useful for hooking down branches of millet-seed from the upper shelves. There was a fore-court in front of the shop where samples of the goods within were displayed on two tables resting on trestles. Amongst the things on view were tanks and bowls of goldfish, a tray of rather lethargic-looking tortoises, carefully arranged pyramids of dog and cat food tins, and also feeding and drinking bowls, not to mention, of course, Sambo and Fred the toucan. The last two (according to their owners) were worth hundreds of pounds a year as crowd-pullers.

At the back of the shop was a long passage, also lined with shelves, which led into a fairly large yard, apparently surrounded by pre-fabs, which in those days were still much used as living quarters by local families who had been bombed out of their own homes more than fifteen years previously. I noticed the door into the yard with relief. When the protection racketeers called for their hand-out I would be able to make a swift if undignified exit through the back premises and away over the adjoining roof-tops. There were three largish rooms upstairs, one of which would make an ideal bed-sitting-room for myself and Tally, whilst another, right in the front, I thought could be turned into a pleasant sunny mews for Random when she was not enjoying the salubrious air in the yard below.

At that time, my grandmother, who was far advanced in her nineties, still lived 'up west' or, more precisely, had a large and luxurious flat at Wilbraham Place, just off Sloane Street, so that I had a refuge to retreat to if things proved unbearable.

Before leaving for Kent and the security of the country, I agreed to spend a day or so at the shop, watching proceedings and picking up a few tips on how to run the place with some measure of success and the minimum of embarrassment to myself and my potential customers.

As I lay in bed that night listening to the quavering, tremulous hoots and sharp, explosive 'keewits' of our local tawny owls and to the distant nostalgic whistle of a train puffing its way across the Weald far below, I had serious doubts as to the wisdom of what I was now morally, if not yet financially, committed to. I loved the country and everything connected with it and I detested any town larger than Rye or Lewes, which were themselves very much a part of the country

which surrounded them. Now I was about to become involved with one of the largest, noisiest and most populous cities in the world! My family and friends were loud in their advice and criticism, telling me, as if I wasn't already painfully aware of the fact, that I was a complete idiot even to consider such a proposition. I was bound to fail and would probably into the bargain be knocked on the head and robbed by a gang of Teddy boys, a form of British wild life which at that time was very much in evidence and making a confounded nuisance of itself in the process.

I had myself thought a good deal about these disadvantages, but some possibly masochistic streak in my nature, coupled with a certain amount of Anglo-Saxon obstinacy, made me all the more determined to go through with it. Be that as it may, a few days later I returned to the shop to serve my probation, my self-imposed unpaid apprenticeship. I was introduced officially to Sambo, who didn't want to know me and made this abundantly clear when I tried to detach him from his owner, about whose neck his arms were twined like those of a small, shy boy. When I persisted, he shook his head firmly, crying pathetically, and clung the tighter, while he pushed me away with one imperious, little black hand. He made no attempt to bite me, as a lesser monkey would have done; as a matter of fact, in all the time I knew him he was never known to bite anyone, even under considerable provocation.

Within a few minutes I had donned a white coat and, though feeling more than a little self-conscious, was properly arrayed and ready for the fray. I helped to carry out the stock and arrange the display tables, which, I was given to understand, was more than half the art of selling. I made pyramids of bogus coral aquarium ornaments. I hung up festoons of millet sprays and I contrived a beguiling battery of minute plastic and Perspex aquariums, each containing two infant goldfish. These tanks, complete with rows of knobs, were designed to resemble television sets, which at that time had only just begun their devastating sweep across the British Isles.

Sambo, when he wasn't happily wrecking the shop, lived in a large heated cage complete with swings, a sleeping platform and a hot water bottle, which had to be kept permanently at the right temperature and which he would carry around clutched to his bosom like a doll. The toucan, Fred, was immured in an ordinary, old-fashioned, wire parrot-cage, which was, in my opinion, thoroughly unsuitable for an active bird of his size and temperament. I made a private resolve to transfer him to much roomier accommodation as soon as he became my legal property.

The shop soon began to fill up and in less than an hour I personally had sold a hamster and cage, two tortoises, a young budgerigar, and several bags of cats' toilet requisites and began to feel a bit more confident in my ability to take my place amongst the princes of commerce. The whole thing was simplicity itself – why on earth hadn't I thought of doing something like this before? The contents of the till mounted satisfactorily and it was only afterwards that it occurred to me as a bit odd that the same customers should return time and time again for different items which changed hands with remarkably little persuasion from me or the proprietors. Days later I found hanging on a hook in the dusky passage a roller canary which I could have sworn I had sold the previous afternoon.

I worked in the shop for several days before I decided that I had learned enough to be able to take the place over and run it with some measure of success. I found, rather to my surprise, that I got on famously with the local inhabitants, some of whom I was beginning to know quite well. When they started to call me 'Jerry', 'Colonel', 'Me old flower' or even 'Viscount' (to rhyme with 'discount'), I knew that I had been accepted. I found that I was becoming interested in the actual running of the business and not only because any sort of transaction in which money changes hands in the right direction is bound to have some appeal. I also enjoyed meeting the cockneys in bulk. During the war I had known and made friends with several cockney soldiers and had been impressed and amused by their irrepressible wit and good humour. Now I was meeting them at home and on their own ground and I had no reason to alter my opinion or to lose my affection for them.

After a good deal of boring and largely unintelligible legal mumbo-jumbo the lease passed into my hands and the shop was mine to make or mar. All that remained before the keys were finally handed over to me was the final stocktaking, which lasted from 07.30 in the morning until long after the other shopkeepers around me had locked up and gone to bed. The result was that I was the better off by some five hundred goldfish, twenty-five tortoises, fifty budgerigars and an aviary containing umpteen small foreign birds. I had also acquired dozens of hamsters, guinea-pigs and rabbits, one Humboldt's woolly monkey, a sulphur-breasted toucan, and a vast number of cages, fish tanks and dog-leads, not to mention mountains of bird seed with esoteric names, such as 'Red Rape', 'Niger', or 'Mazarin Canary'. I was now in business with a vengeance. That evening I walked to the Underground station at the Elephant and Castle, averting my eyes from a long, stationary line of London buses which, in length and colour, reminded me horribly of my bank account.

The two ex-proprietors had agreed, for a consideration, to remain on the premises to feed and care for the livestock until such time as I could transport myself, lock, stock and barrel, or rather, owl, dog and eagle, to my place of gainful toil. On second thoughts I decided that one large, lethal and generally rather overpowering bird of prey was likely to prove quite enough for the nerves of the excellent denizens of Bermondsey. With considerable regret, I found a home for Cheetah, the great horned owl; a home where, after a long period of readjustment, he would be able to continue his crepuscular forays, where he could still winnow his way on moth-silent wings about the sleeping shrubberies on his interminable quest for voles, wood-mice and other small creatures of the dusk.

The village carpenter, an old-fashioned craftsman of a type which I fear is rapidly becoming as rare in Britain as the marsh harrier, made me a splendid eagle carrier for transporting Random on her expedition to the metropolis. This container was a minor work of art, shaped like a miniature horsebox without wheels, with ventilation holes neatly spaced in the front, just enough to let in air but not too much light to make her restless. It was long and high enough for her to travel in comfort but too narrow to allow her to turn round and damage her plumage. The door at one end ran on grooves and all I had to do was to back her carefully into it, making sure that her great wedge of a tail did not brush against the sides and damage the tips of her feathers, a hazard which always has to be anticipated and guarded against when moving birds of prey from one place to another. After one or two trials she got the hang of the thing and would allow herself to be eased into this rather coffin-like container without any attempt to resist. She has always been a first-rate traveller and on those occasions when she does not travel proudly perching on her block in the back of the car she accepts things as they are and settles down in her box to doze the miles away, showing no unseemly haste to emerge at journey's end.

My conscience did not smite me unduly at the thought of uprooting Tally and Random from the country that they loved and from the way of life to which they had become accustomed. I knew that I would have every week-end free to devote to them and I also had a shrewd, uncomfortable feeling that the whole venture was to prove of a strictly temporary nature. Besides, Random was, as I have stated earlier, at this stage of her life lazy to the point of lethargy and needed the strongest inducement to persuade her to fly a hundred yards. As for Tally, he was a dog, and, like all good dogs, he knew that his place was with his master and his happiness came from shared experiences.

There was a large leafy park nearby, which, like so many English town parks, was in reality a piece of enclosed, well-disciplined countryside, full of unexpected birds and beasts. There were newts breeding in the shallow, weedy ponds and hedgehogs which slept away the daylight hours snug in nests of dry leaves deep in the interior of the shrubberies, from which they emerged at dusk, trundling across the well-swept paths, to slip between the palings of the surrounding houses. There they were fed on bread and milk by the kindly inhabitants, who despite their rather unpromising surroundings, were in many cases confirmed countrymen at heart, even if the only time they saw the country was on their annual hop-picking expeditions into the Weald of Kent that I had just left behind.

Chapter Seven

I have always found travelling with livestock, particularly large and conspicuous livestock, to be a bit embarrassing. The much-vaunted British love of animals is not always as apparent as one might be led to believe, despite all the newspaper articles and television programmes devoted to the subject. Quite understandably, not every bus-conductor welcomes a dog on the upper deck of his great lumbering machine, where he holds (and seldom hides the fact that he holds) the same sort of authority as the captain of a man-of-war. Taxi-drivers are inclined to regard potential canine fares with some reserve and to charge accordingly. This sad state of affairs seems to have developed noticeably during the last two decades or thereabouts. During and just after the war I used to travel extensively, generally accompanied by my lurcher bitch Bracken and my kestrel Cressida. Very seldom indeed did we receive anything but the warmest of welcomes, possibly due to the fact that in those far-off days the British took an affectionate interest in what they rightly considered to be a harmless eccentric. Now, alas, nobody appears to have time to regard anything with affection and warmth.

However, at the time of which I am writing, things were still pleasantly casual, as I found when I arrived at Charing Cross Station with Random in her travelling box and Tally smart and debonair with his brand-new scarlet collar and lead. The combined weight of Random and her box nearly crippled me and my progress down the platform, which seemed about ten miles long, must have been similar to that of a condemned man on his way to his place of execution. I was feeling just about as cheerful, particularly as I had overlooked the formality of buying a ticket for the invisible Random and wondered what the reaction of the ticket-collector would be, should he query the contents of the extraordinary and only too visible burden under which I was labouring. I had no idea how Random would behave if I were obliged to extract her from her resting-place into the sudden glare of daylight. What if her jesses slipped from my grasp? The idea was unthinkable. However, I need not have worried. The ticket-collector barely glanced at the box but merely asked for the dog ticket, which of course I couldn't find until I had turned out every one of my seven pockets.

I stumbled onwards towards the line of waiting taxis. Here I was in luck. The driver clambered down to help me. He patted Tally, which was a good sign, and eyed the box, the occupant of which, bored and made anxious by all the unaccustomed din and movement, was now beginning to thunder about in her confined space.

'What you got there, guv? A baby elephant?' he inquired waggishly.

'No,' I replied with as much dignity as I could muster, 'actually, it is a golden eagle – a trained golden eagle.'

'Blimey! Could I 'ave a butcher's?' asked the now fascinated driver.

Feeling that much depended upon the outcome, I eased open the lid. Random, not a feather out of place and looking, if anything, bigger and more beautiful than ever in that unlikely place, stepped out and climbed on to my fist, looking round with complete composure as she roused her feathers and settled them back into place until they assumed once more their state of almost metallic smoothness. I raised my arm so that she had to stretch her seven-foot wing-span, whilst the driver and his mates looked on in silent rapture. I knew then that I had made a friend for as long as I wanted one and, as it happened, from that day forward I was never at a loss for transport. Random, her effort for public relations finished, stepped quietly backwards into her box, which was now resting on the floor of the cab. I climbed in and Tally jumped up beside me. Soon we were speeding south-eastwards through the bustling early morning streets towards Lambeth and beyond.

When we arrived the driver, who had been chatting amiably all the way, helped me carry the box into the shop. I paid what was marked up on the meter. Nothing appeared in the space marked animals or luggage and I mentioned this to the driver.

'That's all right, guv,' he said, 'but I wonder if you could let me have a photo of Random for my little girl. She'd never believe me if I told her I'd been driving a golden eagle about London.'

It so happened that I had a very good photograph in my wallet, one of the first ever taken of her. I gave it to him and he stuck it in his jacket pocket.

'If you ever need a cab, guv,' he said, 'just ring this number' – and he handed me a card.

I took him at his word and he never failed me. I can only guess at the number of miles we must have travelled together, becoming very good friends in the process.

With the help of Charlie, the ancient assistant, whom I had inherited with the shop, I lugged Random's box up the narrow winding stair-case into the front room. I opened it and out she came, to the amazement of Charlie, who was even more dumbfounded than the taxi-driver had been. I up-ended her box, which made a splendid temporary perch, and she flew up on to it, spreading her wings to the sunshine which was flooding into the room through the slightly open windows. I was greeted by Fagin and Sikes, who were waiting none too patiently for the remittance I had promised them. After cups of tea, paid for by me and interspersed with crude cockney witticisms, they departed with scarce a glance at the unhappy Sambo, who was gazing forlornly from his cage. I sent Charlie out for a box of sponge cakes, which, dunked in sweet tea, were, I gathered, amongst the monkey's favourite snacks. He took one from my fingers and ate it absent-mindedly, but it was unflatteringly obvious that he considered me a poor substitute for his two erstwhile mates, who, to their credit, had always treated him with the greatest kindness and consideration.

Before settling down to the stark necessity of earning my living or at least recouping some of the prodigious expenses I had incurred, I took Tally for a walk around the local bomb-sites, which were still much in evidence as monuments to the appalling hammering that the East End had taken during the blitz of late 1940 and early 1941. During the months that followed Tally was to get to know these bomb-sites very well. Although hardly attractive, despite their superficial covering of rose-bay willow-herb, yellow groundsel and ragwort, they made excellent dog-exercising places and had a natural history all their own. I once found a number of elephant hawk-moth caterpillars munching happily amongst the willow-herb and I even saw a pair of black red-starts near one of them and have no doubt that these enchanting little birds were nesting in an inaccessible crevice in the cliff-like walls of some ruined building.

Tally, who had never been in London before, took everything in his stately stride, stepping out a few yards in front of me, the creamy banner of his tail flaunting high over his glossy back, and grinning his acknowledgment of admiring looks and friendly words from complete strangers. He might have been a town dog born and bred.

The bomb-sites were the meeting-places, the gang headquarters, of all the rather raffish, thoroughly independent, but none the less engaging, local dogs. These, although in most cases of almost unidentifiable ancestry, none the less fell into certain well-defined groups. There were small, smooth, black-and-tan ones that were generally described by their optimistic owners as 'Manchester terriers'. There were medium-sized black ones that would probably tell you, if you asked them, that their great grandmothers had been seduced by a labrador. And there were shaggy ones of all colours and sizes whose twinkling eyes peered through a screen of tangled hair and were just honest-to-goodness Dog, with no pretensions to being anything in particular and not caring a damn for anyone's opinion anyway.

These dogs were very much a feature of that part of London at that time. Most of them had homes where they were well-loved members of the family, but as the human members were often either working or at school for much of the day they had between them worked out a routine that appeared to be mutually satisfactory. They would be released in the morning, when they would trot off to the local park or, more probably, to their own particular bomb-site, where they would meet their friends, whom they had probably known for years. Here they would play and scamper or go for little expeditions of their own. It was quite usual to see half a dozen or so of these cheery little devils jogging along together like a group of schoolboys. They interfered with nobody, seldom fought amongst themselves, and, unlike their less fortunate contemporaries in the country, were safe from the temptation to worry sheep or other livestock. Coming probably from generations of town-dwelling dogs, they knew their traffic drill perfectly and never appeared to cause an accident, although of course in those days the quiet back streets held precious little traffic anyway.

When Tally, the haughty stranger from foreign parts, first arrived on the scene, some of the bolder and more impertinent members of these closely-knit clans were inclined to say rude things about his admittedly rather exotic appearance and posh accent. They learned quickly that it was prudent to leave this standoffish canine toff to his own devices.

Although these London dogs had homes and owners to which they were in most cases devoted they had none the less developed certain characteristics which they shared with their semi-wild contemporaries, the pariah dogs of the East. Each small pack or group had its own leader, which was obviously the most intelligent and resourceful though not necessarily the largest and the strongest of that particular clan. Each

72

group kept to its own well-marked territory, out of which it did not normally stray and which it guarded against intrusion by other packs. This behaviour had its parallel amongst the human element of the district. There were, for instance, the 'elephant boys', so called not because they affected tusks and a trunk, but from the district (the Elephant and Castle) which was their stronghold and from which they sallied forth to do battle with the Deptford mob or the Lewisham lot. The law of the jungle ruled supreme, though the gangs generally contented themselves with beating up their rivals and did not interfere with neutral passers-by unless, of course, one was unfortunate enough to get in their way.

My first job on returning to the shop after that preliminary reconnaissance of the district was to make a larger cage for the toucan, Fred. I am no carpenter, but after a couple of hours or so, during which wood, wire netting, nails and curses were about equally proportioned, something resembling an indoor aviary appeared along one wall of the shop. About six feet long by four in height, it wasn't much but it was none the less the nearest approach to unrestricted movement that poor old Fred had known for many months. He showed his gratitude by nearly severing one of my thumbs with his wicked double-edged bill whilst being transported to his new home. Once ensconced, he sat looking as morose as ever, with his ridiculous little tail cocked over his back and clattering his great multi-coloured banana-shaped beak at me with the sound of a dyspeptic machine-gun. It took him some time to get the hang of his new quarters and to learn to spring heavily from one end of the cage to the other, landing on each perch with a jarring thump that would bring mountains of tinned cat-food crashing to the floor and render Charlie almost incoherent with rage. Later, Fred became quite tame, gave up biting, and could be carried perched on my hand for periods of exercise in an empty room. However, being a fruit eater, and his droppings therefore both copious and unpleasant, it was imperative to

73

cover the floor from wall to wall with layers of newspaper before his liberation.

Sambo, the woolly monkey, wasn't exactly house-trained either, but what he left behind was small and dry and much easier to cope with, possibly because of the high proportion of cereal in his diet. Sambo himself, after a suitable period of grieving for his ex-owners, came at last to realize that they would not be coming back and decided that I was not too bad a substitute after all. I would find him on my arrival at the shop each morning waiting for me by the door of his cage. I would pull back the bolt and he would reach out his arms to me and swing up on to my shoulder. Charlie, who was already there, would go across the road to the café for a jug of tea and we would all three take part in our daily ritual. Sambo was passionately fond of tea but none the less, having finished his cup, would politely hand it back to me, shaking his head if offered a refill. His breakfast consisted of half a banana, two or three dates, and a small plate of corn-flakes with milk and glucose, which he ate quietly and contemplatively before watching us prepare the shop for the day's work.

He wore a woolly jersey, one of a number which had been knitted for him by a local admirer who used to come each week to take away the soiled one he had been wearing and to bring a clean one which she had washed for him, so that he was always tidy and well-dressed. During his periods of freedom, which in fine weather lasted most of the day, he was not hampered by a chain or even by a leather lead and yet he did not take advantage of his liberty by shinning up the adjoining roofs or raiding the local stalls, as an Indian rhesus or African vervet would almost certainly have done.

Sambo had been taught (or had taught himself) one most intriguing party piece. He used to spend a lot of time playing about the fore-court, gambolling between the table legs or just sitting on his haunches in the sun, apparently lost in thought, but, no matter where he was or what he might be doing, if Charlie or I shouted 'Quick, Sambo, here comes a copper!'

74

he would rush up, scampering on all fours, hurl himself into one's arms and bury his face in one's chest before looking round grinning and chattering, as if to say 'I'm all right here, mate!' Goodness knows where he had learned this trick but it seldom failed to bring rounds of applause from the crowds of onlookers who always seem to haunt the vicinity of a pet-shop.

Random for her part took to London life like the proverbial duck to water, which shows just how resilient she is. I fixed two perches at opposite ends of the room that ran the length of the upper storey and here she reigned supreme. The sun, when there was any in evidence, poured into the room, and there she held court, greeting her many visitors with an old-world courtesy, reminiscent of the lady of the manor receiving a deputation of villagers. There was nothing but thin glass between her and the low, flat roof-tops of south London, which stretched away mile upon mile until lost to view amidst the smoky heights of Blackheath, which stood between her and the distant Weald of Kent where she had been reared to majestic maturity. Any lesser bird of prey, such as a goshawk or sparrow-hawk, would have crashed straight into the glass, probably killing itself in the process, but not Random, who sat with one foot tucked comfortably into her breast feathers – an enviable picture of pure relaxation – turning her head to watch with interest the passing of people and dogs in the noisy, thronging street below.

I cleared a mass of lumber, broken-down hutches and other impedimenta out of the yard and installed a huge, broad-based log, about three feet high, which I received with the blessings of a local timber yard. I drove an outsize staple into one side a few inches from the top and fixed a length of strong clothes-line to it. On fine days I would carry Random down the creaking stairway out into the yard and fasten her to this splendid block. Here she would perch contentedly, watching the sky and noting the passage of every pigeon that crossed her line of sight. Although she seemed perfectly content in these somewhat unpromising surroundings I was not entirely

75

happy about her. She might not have wanted exercise but she certainly needed it.

I had taken Tally to the local park several times and I knew just what the lay-out was. It might not be ideal but I had, on occasion, flown Random under even less auspicious conditions in the country. I had read the formidable list of regulations and prohibited activities, which included such improbable occupations as whippet-racing, grazing swine and asses, playing games of chance, or even reclining on the benches in a verminous condition, but no mention whatsoever was made about flying eagles to the danger of the public.

The park was presided over by three keepers, who wore a brown, tweedy uniform that for some reason made them look like male equivalents of the Women's Land Army, and were quite differently arrayed to their counterparts in the Royal or Central Parks, who, in their conventional navy-blue serge, could have been mistaken for prison warders or even itinerant RSPCA inspectors.

I rose early and with Random hooded and statuesque on my fist stole out through the side door of the shop and into the silent, deserted streets. I dreaded the whole operation as I had a horror of attracting undue attention, although I doubted if the local inhabitants, who had witnessed all that the Luftwaffe at its most unfriendly could throw at them, would be unduly perturbed at being confronted with twelve and a half pounds of fighting fury.

The park was only about a quarter of a mile from the shop but the journey seemed interminable. Already a few itinerant stall-holders were manœuvring their cumbersome tarpaulin-covered barrows into position, but they hardly glanced at me, nor, strange as it may seem, did they appear to notice Random, perched, still as a museum specimen, upon my fist. I took heart and hurried onwards, Random's weight growing ever more burdensome and her grip, which normally I hardly noticed, almost paralysing my left forearm.

A bunch of noisy schoolboys dribbling a huge yellow ball

materialized apparently from nowhere and I crossed to the opposite side of the road, hoping to slip by undetected. Alas, it was not to be. One of them, gazing at the motionless, hooded figure of Random, shouted out 'Coo, look – a polly what's got its 'ead in a bag!' Almost choking with mixed rage and amusement I strode on, dreading the sound of following feet, but luckily I was soon hidden behind a half-shattered wall that skirted the alley-way that led up to the park gates. At this early hour the gates stood locked and forbidding, festooned with padlocks and papered with notices defying anyone to do just about anything on pain of fearful penalties unless granted permission in writing by the Clerk of Works, whoever he might be. However, large holes had been cut at convenient intervals in the wire which guarded this paradise, presumably by rebellious youngsters who saw no reason why their courting forays should be curtailed by the dictates of an unromantic and featureless authority. These gaps suited me admirably and after a certain amount of commando-like scrambling and wriggling I managed to ease myself through with the indignant Random, who, being hooded, couldn't see what was going on and feared the worst. Having regained my feet I strode as nonchalantly as possible forward into the deserted waste of muddy grass.

An earlier reconaissance had established that the park did not open officially until half past eight and, it being now just on seven o'clock, I had an hour or more at my disposal, free, I hoped, from the prying eye and rasping voice of petty authority. With Random on my fist I felt huge and strong and ready for anything, much as some wrong-doers feel, so I understand, when armed with gun or flick-knife.

Conveniently placed in the centre of this unattractive wilderness stood a pair of goal-posts, ideal launching-pads for Random to take off from, and with no trees close by in which she might be tempted to take stand and contemplate the situation. I unhooded her and she bated like a fury, nearly swinging me off my feet with her wild plunges. I knew that

77

this was just to show how fit she felt and that she was in the mood to take on anything that needed attending to. I swung her back on to my fist and she sat there panting and gripping hard while she stared round at the uncompromising landscape about her. At this time she was at the peak of condition and obedience, without a trace of the sometimes rather embarrassing independence of spirit which she would gain with maturity. I slipped the leash from her jesses and raised my left arm. With hardly a movement of her wings she floated on to the bar of the nearest goal-post, swinging round at once to face me, eager for whatever was forthcoming.

I had brought her well-loved, much-mangled lure with me and this was now garnished with the leg of a white rabbit which I had bought from a poulterer the day before. If there was one form of meat which she loved above all others it was rabbit, adorned with a satisfactory wad of fur to titillate her far from jaded appetite. I walked back a few yards and Random began to yelp with her curious half-barking, half-mewing call – a call that had a strange resemblance to that of a herring-gull but with a hint of hidden menace in it that is entirely her own. I did not want any complications on this her first free flight in London and so I threw out the lure when I was only about a hundred yards from where she perched, watching me, her wings half opened and with dancing yellow feet, keyed up like a high diver about to take the plunge. She came at once in one long slanting glide and I gave her a small reward to show her how pleased I was.

I had a struggle to take the lure away and replace her on her perch but I wanted to make certain that her first faultless approach could be followed by an encore worthy of the occasion. She had got the hang of the thing now, knew what was expected of her, and came the full length of the deserted playing-field, landing with the precision of a homing bomber. I gave her a good feed before calling her from the ground to my fist for a last piece of fresh liver. It was a curious, exhilarating experience to see her come hurtling in to me across that mist-shrouded autumn park, her great wings outlined against the cranes and derricks of the distant Surrey docks. I had pushed my luck far enough and so I hooded her up, struggled once more through the gap and crossed the old bomb-site which divided the park from the shop. All had gone unexpectedly well and a golden eagle had flown free and happily in London five years before the much over-publicized escape of the unfortunate refugee from the Zoological Gardens in Regent's Park.

Chapter Eight

The fact that a large, tame and exceedingly handsome eagle could be seen and admired in S.E.17 quickly came to the attention of the local press and reporters were soon ringing up and asking for interviews. It occurred to me that a certain amount of sympathetic and respectful publicity would do neither her, the shop, nor myself, any harm, and I decided that the sooner she was given some sort of official status the better. Therefore, as the press photographers wanted to take pictures and as the best place to take these pictures was outside against the background of trees, however shabby, the co-operation of the powers that ruled the park was sought, and willingly given. Random soon became used to hordes of admiring urchins and their elders and I am glad to relate that during our entire stay in London nothing untoward occurred, although I must admit that I had one or two anxious moments. One result of all this was that Random was more or less given the freedom of the park and, possibly impressed by the honour that had been bestowed upon her, was always on her best behaviour. People would produce cameras from nowhere and it intrigues me to think that there must be hundreds of photographs of her resting forgotten in albums all over south-east London and beyond.

Another by-product of her increasing fame was that we were asked to appear in person in support of such deserving causes as fêtes to raise money for repairs to the local Scout or Guide headquarters, or perhaps for comforts for the old-age pensioners or the blind. One way and another Random certainly did her bit to help community life in the Borough, as it was universally known to its inhabitants and which was in those days very much of a village, in spirit if not in fact, and surrounded by other 'villages', all with their own individual character and peculiarities.

Tally too, now completely re-adjusted after his release from quarantine, was perfectly at home in London. Tally, in fact, was perfectly at home anywhere – he was that sort of dog. He would thread his way confidently amongst the hurrying crowds, head and floating tail-plume held high, and when released in the park would trot off happily about his own affairs, interfering with no one and leering happily when spoken to or patted by his admirers, most of whom, to Tally's secret indignation, mistook him for an Afghan hound. This was an understandable case of mistaken identity, because he was almost certainly the first saluki to have appeared in that part of the world. He was particularly popular with little girls, immense numbers of whom appeared whenever he trotted on to the scene and formed what can best be described as Tally's fan club. They were fascinated by his Arab name, asked me to repeat it in full many times and never seemed to tire of hearing it. It was indeed an intriguing name: Talarah El Fidaui Ibn El Gazaal.

More especially did we enjoy exploring the deserted docks and wharves, late in the evening when the men had gone home and the silent cranes reared their stark heads against the encroaching gloom of a November dusk. Lights flickered across the river from Limehouse Reach and from the grey swirling Thames came the shouts of lightermen and the deep-throated call of sirens as the ships swung out from Butler's Wharf or Chambers' Wharf on their slow journey to the open sea, still many miles down river to the east. Leaning on the breeze or scavenging about the foreshore were the gulls – herring, black-headed, and more rarely that huge, savage but handsome freebooter, the great black-backed gull, with its six-foot spread of sail, gruff barking voice and pitiless, ivory-yellow eye. The sight of these magnificent marauders always brought back memories of some remote Western stack, lapped by the swelling Atlantic, pink-flecked with patches of sea thrift and clamorous with the calls of kittiwakes and the wild musical voices of choughs – most enchanting of all the crow family.

Scattered about the narrow streets just behind the dockland were numerous tiny pubs, the rallying point of seamen of all nationalities but more particularly of the Scandinavian races – Swedes, Norwegians and Finns. Tally and I used often to call in for a pint and to hear news of Mombasa, Dar-es-Salaam and Mogadishu and to listen once more to the language of the Kiellands, my old friends from Tanganyika, who, so I had heard, had now returned to Oslo.

Owning both a dog and a pet-shop it was only a matter of time before I became friendly with most of the other local dog-owners and I suppose it was inevitable that I would be asked to look after some of these dogs when their owners went on holiday. Rather reluctantly I agreed to take on one or two of those belonging to my closer acquaintances. I did not own a boarding-kennels, nor had I ever had the slightest inclination to do so. For years I had lived with one or very exceptionally two dogs and that was the way I wanted things to continue. Alas, it was not to be. In my folly I found it hard to refuse when some brawny docker presented himself before me with a hairy mongrel trotting confidently into the shop beside him, wearing a brand-new collar and lead bought, or so I hoped, from me, and stood looking hopefully while Butch or Blackie grinned cheerfully up at me, obviously expecting a welcome. The net result of all this was that I had to move Random and Tally into the two spare rooms in order to make room for a veritable avalanche of dogs.

This more or less enforced amateur dog-boarding was an interesting experience. I became the temporary landlord and foster-father of dogs of all ages, sexes, breeds, and combination of breeds. I made it absolutely clear that while I would look after them to the best of my ability I would not be ultimately responsible for any loss or injury and I made the dogs' owners sign an official-looking document to that effect. All the dogs lived happily together in one large sunny room, each of course with its own bedding, box or basket. At one time I had fourteen dogs and yet not once did I have a serious fight on my hands, nor did any of them abscond in order to look for their own homes. Having installed all my guests they then had to be fed and exercised. I used to boil up an enormous cauldron of meat and gravy. I added biscuit meal by the pailful and let the whole lot cool off in time for the main meal, which was in the evening after I closed the shop. I put down a huge piece of linoleum in the centre of the floor and decanted the whole mass of rich meaty goodness on to it, whilst I stood by in case of a breach of the peace. As their dining table was exceedingly spacious and the meal ample, there was never any trouble, but I must admit I was jolly lucky. There could have been the most appalling chaos.

In a corner of the park stood a hard tennis court, surrounded by a ten-foot fence of stiff wire netting. It was rarely used for the purpose for which it had been designed but it made a heaven-sent exercising ground for obstreperous London mongrels. I used to take them six at a time to this gravel paradise and, after closing the gates and tying them together with my handkerchief, I would let the pack loose to gallop and play. Half an hour of this was all they needed. With heaving flanks and laughing eyes they would come lolloping up to be re-leaded and walked home at a (comparatively) sedate pace. Later, when they had been with me for some days, I let the whole lot loose in the park. By now they knew each other and me well, and so after scouring the muddy apology for a plain and chasing each other until their tongues lolled out they would return in groups of two or three, tired, contented, and more than ready for their evening meal.

One evening when I was rounding up my gang of hooligans after a particularly lively sortie I looked up after snapping on the last lead and saw looming over me an immense docker. At his heels was an equally immense dog, beside which Cerberus would have looked like Mickey Mouse.

'Do you mind dogs, Guv?' he rasped.

'Not at all,' I replied, grabbing Tally, who was about to launch a suicidal attack upon the monster before him, 'actually I am very fond of them.'

'What I mean is, do you look after them when their owners is away, like.'

'Well, I do sometimes,' said I, blanching at the thought of what I felt certain was to follow.

His face broke into a broad grin. I was much relieved that he didn't think I was trying to be funny, because he was just about the toughest-looking individual I have ever encountered. He appeared to be about eight feet tall, with hands which resembled, not so much hams as haunches of beef from a Hereford bull. He had, however, a humorous twinkle in his eyes and I could see that with him as a friend one wouldn't need to worry much about enemies.

'Well, it's like this,' he said. 'My missus 'as got to go into dock for an operation, and as I'm workin' all day there won't be nobody to look after Rusty 'ere. I'd like 'im to go somewhere where 'e would be treated proper, like. I don't want 'im to 'ave to go to one of them kennels where they shuts 'em up all the time.'

Tally had calmed down by now and decided wisely that single combat with this veritable Beachy Head of a dog wasn't quite his line. I looked at Rusty, who was sitting on his haunches grinning, his tongue lolling and showing a set of teeth of which a Bengal tiger might well have been proud. It was obvious from his smile that he regarded the whole world and everything in it as his friend. I took a good look at him as he sat there on his haunches. He was colossal — a cross between a Great Dane and a Boxer, with almost the height of the one and the muscular athletic build of the other, magnified in proportion. His tail had been docked and what remained resembled the stump of a well-grown oak tree. In colour he was a lovely, rich, mahogany brindle with a black face like a mask from which honest, kindly eyes smiled out at the world about him. I went up and tickled his ears and he gave me a great splashy kiss on my hand, which finally made up my mind.

'All right,' I said, 'I'll take care of him for you. Bring him round to the shop when you are ready. I'm sure he'll be no trouble and settle in well.'

The man's look of relief was worth all the worry and I felt a glow of satisfaction as I watched the two striding happily out of the park together. I have seldom seen a better adjusted pair.

A few evenings later Rusty and his master re-appeared at the shop. After the usual formalities I took charge of the dog whilst his owner, after telling him to behave and to cause no trouble, left to start his shift at the docks. I fastened Rusty by his lead to the counter while I went up to prepare his bed and get him a meal, but when he saw his boss departing, as he no doubt thought, for ever, he leapt forward, towing counter, till, pyramids of displayed goods, and cages of hamsters behind him as if they didn't exist. This sort of thing wasn't part of the deal and so I unhooked Rusty and took him upstairs, where I introduced him to his future bedfellows.

There were five assorted puppies already in residence. Rusty was in fact a puppy himself, being well under eighteen months old. Like most very large dogs who have not reached full maturity, he still had to grow into his skin, which hung around him in folds and made him look rather like a fourteen-year-old boy wearing his father's overcoat. This, with his pathetic, pleading, 'baby dog' look, added greatly to his considerable charm. His character was as pleasant as his appearance was formidable; in no time at all he was beset by the entire puppy pack, who swarmed all over him, chewing his ears and grabbing any bit of him that they could reach. I stood by in case of trouble but I need not have worried. Although he looked a trifle embarrassed at first, he soon shook off his self-consciousness and was rolling and gambolling about all over the room with the grace and elegance of a six-months-old Clydesdale colt.

Because of his size and gentleness he was soon accepted as the natural leader of my assorted pack of temporary guests. This was useful because when we were out exercising in the park they all stuck close to him and, despite the twilit gloom of river fog and factory smoke, his great figure would loom up out of the murk surrounded by his smaller companions like a battle cruiser escorted by a flotilla of playful and noisy destroyers.

Despite his bulk and an air of magisterial gravity (which was utterly misleading) Rusty could jump like an Exmoor stag and swim like an otter. Ignoring the notices threatening the direst penalties he would sail over the horrifying spiked railings surrounding the pond in the centre of the park and, plunging into the green and slimy depths, cleave his way to the solitary island in the middle with all the confidence of an Irish

86

water spaniel. Full of self-satisfaction at his bravado, he would rout about amongst the bushes, snorting and puffing like a hippo and incidentally scaring the daylights out of the mallards and moorhens who were the official denizens of the place, before slipping once more into the water for the return trip, grabbing as he passed some piece of floating debris, which he would present to me as a peace offering, his ridiculous stump twiddling and his little laughing eyes looking into mine from the depths of his dusky mask as he did all he could to allay my justifiable wrath. These aquatic expeditions of his were unpopular, both with uniformed authority, who frowned upon these deviations from the path of righteousness, and with myself, who was afraid that one day he would misjudge his leap and impale himself like a chicken on a spit. Eventually, by a suitable mixture of oaths and endearment, I taught him the error of his ways, and we had no further trouble of that sort.

Rusty stayed with me throughout that first summer in London and was still there when autumn yellowed the leaves of the sooty-barked, melancholy plane trees and brought with it its promise of hot chestnuts and crumpets oozing with butter.

It was my practice each Saturday evening to tip the week's takings (often considerable) into an old brief-case and to take them to my grandmother's flat, from which I would transfer them to the bank on the following Monday morning. If the weather was pleasant, with that hint of excitement in the air that seems to me to belong to autumn alone of all the seasons, I would walk the whole way, closely flanked by Rusty and Tally, who by now had learned to accept and tolerate if not actually to love each other. The journey led us via the Elephant and Castle over Westminster Bridge, through St James's, with its lake now thronged with tufted duck and occasionally with more exciting winter visitors, into Green Park and finally, threading the victory monument, to Hyde Park itself. The first part of this odyssey was a cheery progress. "Ullo

Gerry, me old flower!' the stallholders would bellow as I passed. 'Tiking the blood 'ounds 'ome to the fam'ly estite? Wotch aht the elephant boys don't getcher!'

'Never mind the elephant boys,' I would reply, patting the heads that were jogging along nearly level with my waist, 'these blokes haven't had their supper yet!'

Actually, I wasn't at all certain how either of the dogs would react in the face of crisis. I was soon to find out. One evening, tired, with an exceptionally heavy brief-case, (it is surprising how much twenty or thirty pounds in silver and copper can weigh) I reached the southern boundary of Hyde Park. I decided to dawdle for a spell and to rest my aching arms, which by now seemed to have reached almost orang-outang proportions, due to the strain inflicted on them, before continuing the last leg of the journey to Sloane Square.

Darkness had closed in as I crossed the Row to the west of Hyde Park Corner and looked for an unoccupied bench on which to relax and gain my second wind. I found one and sat down, propping my brief-case beside me. Tally and Rusty lay at my feet, unseen, only their deep sighs and quiet regular breathing announcing their confidence-inspiring presence. I thought at first that the park was deserted but this was not so. Groups of whispering figures appeared to lurk close to the trunks of every tree. I lit a cigarette and drew a deep breath. Somehow I could sense, rather than see, a group detach itself from the deeper shadows and move towards me. I felt much as Mole and Rat must have felt when benighted in the depths of the Wild Wood.

I stood up, gripping my brief-case. The two dogs rose silently and pressed against my legs. I reached out and felt the powerful shoulders of Rusty on my left. I could feel the hackles lifting under my hand, an almost soundless growl vibrating the muscles of his throat. Tally on my right was tense and alert, his white-tipped stern rigid as a flag-pole. I strode forward towards the lights that were now twinkling in the distance — too great a distance for my peace of mind. I could hear the scuffle of feet behind me as I lengthened my stride. From right and left other figures appeared to be closing in on us along the tree-lined avenue through which we had perforce to travel before reaching the well-lit security of the Brompton Road. London at that time was in the grip of a crime wave which almost equalled that of America. Hyde Park, I felt, was little safer than Central Park, New York.

A knot of hatted figures appeared suddenly before us, waiting, it seemed, for my approach. I gave one sharp word of command. Rusty, with a leonine roar that shattered the night, plunged forward into the darkness with Tally close behind him. All around me I could hear running feet rapidly withdrawing, followed by the pounding gallop of heavy bodies as the two dogs, drawing a crescent round me, sped these would-be felons on their way. Within seconds both dogs had returned, panting and grinning, and I knew that I would have little further to worry about on that particular journey. But it was a well-needed warning to avoid the erstwhile peaceful London parks after nightfall.

This incident increased my respect for dogs in general and these two in particular and furthermore helped to reinforce a belief I have always held that a good dog will almost invariably rise supremely to the occasion and is in fact often far more in control of the situation than its owner. Rusty at this time was still young, virtually a puppy. He had no streak of viciousness in his nature. Yet the guarding instincts of the two breeds from which he sprang were lying there just below the surface, ready to awake to a challenge triggered off by the sort of circumstances that had just arisen. Tally, being a saluki, was a hunting dog pure and simple and not by tradition a guardian of home or property and yet he was quick to follow the lead provided and to act accordingly.

Speaking to a policeman acquaintance of mine a few days after this rather unpleasant happening, I found that he took the matter very seriously indeed — in fact, he more or less told me that only an idiot would venture into the park after darkness had fallen. He told me furthermore that numerous pilfered brief-cases and handbags had been found discarded amongst the shrubberies which are such a feature of the place – what had happened to the owners of these things he either didn't know or didn't wish to relate. Apparently this sinister reputation, for some curious reason, applied only to Hyde Park; the other two, St James's and Green Park, being almost as innocent as kindergarten playgrounds.

A somewhat similar incident occurred not long after this, an incident in which the redoubtable Random was to play the heroine, a role for which she seems eminently suited and which she thoroughly enjoys. I had decided to take her and the two dogs down to Sussex on Sunday to enjoy a spell of real freedom in the country that she and Tally knew so well but had not seen for a considerable time. I was by now becoming disenchanted with London, the shop and my general way of life, and was feeling morose, edgy and generally bloody-minded. I knew that the view from the summit of Firle Beacon, with its inimitable mixture of sea air, thyme and sky-lark song, would do much to restore me and make me whole again. With that in mind I left the two dogs on Friday night in the care of 'Nanny' Hill, my grandmother's housekeeper, friend and confidante for the past fifty years or more.

Somehow, helped by Charlie, my troglodytish assistant, I struggled through the work on the Saturday, thinking only of the morrow. I had no dogs other than Rusty lodging with me at the time and Charlie had proved himself trustworthy and capable enough to feed and water Sambo the monkey, the toucan and the few dozen hamsters and cage birds that I had in stock at that time. He had been coming in every Sunday morning since the shop had opened and could cope with almost any eventuality that might arise – except Random, he drew the line at her, for which he could hardly be blamed.

Random had become accustomed to travelling between the shop and the flat in a large oblong cardboard box which had originally contained my grandmother's television set. This box was strong but light, which, as Random herself weighed nearly thirteen pounds at the time, was just as well. It was about three feet long and approximately eighteen inches high, with a piece of carpeting inside to give her feet a grip. Provided Random lay down as she was wont to do, she could travel the comparatively short distances involved in perfect comfort. As long as she felt the movement of being transported, either by bus, Underground or hand, she would remain passive and relaxed, but if the box was put down on the ground for more than a few seconds she would assume that we had reached journey's end and unless I was careful she would barge out through the opening, which was secured, or intended to be secured, somewhat unprofessionally, by a pair of shoe-laces. The bus-conductors on the route between the shop and the Underground station knew us both well by now and would somehow manage to find room for Random's mobile mews in the space usually reserved for prams, push-chairs, shopping-baskets and so on. I used to ponder with some amusement as to what the reaction of the respectable cockney housewives would have been had they known that the potential terror of the skies was lurking with all her latent power concealed only by a thin layer of frail cardboard within a few inches of their fashionably shod feet.

On this auspicious Saturday evening I heaved the eagle-filled ex-television-set-container off the platform of the lower deck of the bus and strode across the bomb-site which stood between me and the steps leading to the warm, smelly, thronged inferno, which, after thirty or so jolting and noisy minutes, would disgorge us, limp but happy, at Sloane Square. As I crossed the bomb-site, tripping in my haste over pieces of crumbling masonry, I saw that I was not the only occupant of this unattractive place. There must have been nearly a dozen youths of a species easily identifiable at that time but now extinct for twenty years or so – drain-pipe trousers, winkle-picker shoes and bootlace ties: there was no mistaking the plumage. Whether known as Edwardians, Teddy Boys, or just plain Teds, the animals were the same.

Seeing a potential victim, laden and apparently staggering under the weight of what they obviously took to be a comparatively newly-invented television set, they advanced upon me *en masse* and began to take what, for want of a better word, had better be described as the 'Mickey'. The apparent leader of this unsavoury pack was a tall, consumptive-looking lad in his early twenties, who, in the murky glow of the distant light, seemed to be all arms and legs and about as muscular as the skinned rabbit that I had just bought for Random's supper. He would not have been a particularly formidable proposition on his own — but unfortunately he was not on his own. His companions, hoping for a bit of sport at the expense of the solitary stranger, closed in on me from all sides. I didn't like it one bit. 'So we've gorn and bought a nice new telly, 'ave we? Let's 'ave a butcher's, eh?' I made what I hoped would pass for a conciliating chuckle and continued on my way, praying that they would lose interest and clear off. ''Ere, you! I'm talking to you,' said the leader, who looked exactly like an anglicized version of a bit-part actor from the cast of *West Side Story*. I was beginning to get angry. My temper isn't particularly placid at the best of times and I object to being bullied. I was extremely fit at that time and I could use my fists fairly effectively in an emergency; also I knew a good deal about unarmed combat, at which I had been reasonably proficient in the army.

I stopped, faced my persecutors, and having placed the alleged television set, the cause of the contention, on the ground in front of me, said loudly, emphatically and aggressively, 'If it's any of your bloody business, I've got a golden eagle in this box — and a bloody dangerous one at that! Now sling your hook, as I happen to have a train to catch.' The fellow in front of me stared in amazement, thinking no doubt, that he was dealing with a madman — a new and startling experience for him. At any rate, he paused and remained silent.

Random had heard my voice. Her box was stationary and she probably thought that she had arrived. I saw the frail cardboard vibrate slightly as she tensed her muscles before charging through the narrow opening of the box, bursting into that dimly lit bomb-site with all the impact of a fighting bull into the arena. For a second she paused as her greeny-gold eyes took in the situation. Her hackles rose and her flailing seven-foot wings shot out as she went for the leader of the gang with her swift, shuffling run, her head thrust forward truculently. She was gaining speed, had in fact risen a few feet from the ground, when I hurled myself forward and grabbed her leash, bringing her floundering to an undignified halt within inches of her nearest foe.

Random in a temper is an alarming sight, even to those who have known her for years, as my wife Imogen will confirm. The effect on these young hoodlums was almost ludicrous. For seconds they stood motionless as pieces of uncouth statuary, in the exact attitudes they had struck before this apparition burst upon them. In those days television had not made the impact that it has today, when one can hardly turn the thing on without being confronted by eagle, hawk, leopard, cheetah or some other rapacious denizen of the wild. The very word 'eagle' held an aura of mystery, of dauntless courage and untameable ferocity. It certainly had a most satisfactory effect in this case. As I persuaded Random, who was straining at the end of her leash and leaping up and down in a frantic attempt to close with her enemy, to step on to my fist, where she still stood mantling and threatening her demoralized prey, the gang of bully-boys stood not upon the order of their going but swiftly melted away into the night.

Entirely unperturbed by this episode, Random put on a superb display the following day, launching serenely from the summit of Mount Caburn, fingering the channel breeze with flexible primaries as she rode the rising air currents, sailing in great arcs far above the tiny patchwork fields of the sleeping Weald so far beneath her, yelping loudly with exultation in the

93

freedom of the skies she loved so deeply. Tally and Rusty, happy to feel again the springy, elastic, downland turf with its promise of innumerable hares and the sudden whirring shock of the rising partridge coveys exploding from the spiky stubble, galloped and bounded far out on my flank, clearing gorse-bushes, fences and the sunken, chalky paths in an ecstasy of happiness in their new-found freedom.

Chapter Nine

It was then, lying amongst the tussocky grass looking towards Birling Gap with a distant glimpse of the grey-green channel, that I decided that I must leave London – must give up the pet-shop. I was enjoying the experience in many ways. I had met a lot of interesting and unusual people whom I would otherwise almost certainly never have met. I considered I was doing a good job of work, both for the animals in my charge and, equally important, for their prospective owners. Had it not been for Random and Tally I might well have stayed there for years, becoming more and more involved in local affairs, until I became like the prisoner who, on being released from the Bastille after twenty years' incarceration, returned there voluntarily, having decided that it was the only place where he felt at home in a strange and hostile world. Yet somehow the prospect of spending the next thirty years or more wearing an off-white overall, selling hamsters and gold-fish, and giving advice on how to house-train kittens, had but limited appeal. It didn't seem somehow to be exactly fulfilling my destiny.

Rusty's owner, his wife having recovered from her operation, came to see me, and it was impossible to say which of the two was more delighted at their re-union. Rusty's master, Tug Wilson, had the sort of typically British face that made one feel that he ought to be wearing a stove-pipe hat and a union jack as a waist-coat. His humorous but slightly truculent face lit up with a grin of happiness such as I have seldom seen before or since and was well worth all the responsibility which I had taken. Rusty's equally formidable countenance was one huge pink leer as he almost turned himself inside out with excitement. After the emotion had died down a little, Tug asked me if I could possibly keep Rusty for another week, because he was then going to take his whole family, children, dog and all, to the hopfields of Kent for their annual outing. His wife would not be able to take part in the picking, but she would enjoy the change, benefit from the fresh country air and could keep an eye on Rusty, who would be taking part in the Londoner's traditional annual beano for the first time. Tug was almost inarticulate as he thanked me and said goodbye to Rusty, who obviously thought he was going home. However, he was very fond of me by now. We understood each other and he soon settled down again. We resumed our nocturnal explorations of the silent, sleeping dockland and so the next few days passed happily away.

After Rusty's departure I had no real reason to stay on in Bermondsey. My only responsibility was to Sambo, the woolly monkey. Woolly monkeys are delicate and do not normally do well in captivity, certainly not in the conditions likely to prevail in even the best organized pet-shop. They are exceedingly sensitive to cold and therefore their living quarters should be kept at a temperature of at least 60 degrees Fahrenheit. And so, however tame and companionable they may be — and nothing could have been tamer and more companionable than Sambo — they cannot be taken for long country walks, which in my opinion all real pets should be able to enjoy. They are also rather difficult to feed, needing all sorts of additives and supplements to augment their basic diet of fruit and cereals. The idea of an active, mentally-inquiring animal like Sambo being permanently confined in solitary splendour to a cage became more and more repugnant to me.

It so happened that amongst my large and exotic collection of acquaintances was a man who specialized in South American primates. There was little he didn't know about them and he had even been successful in keeping the rare and still more delicate Red Howler monkey for some years. At this period he owned a black spider monkey and, what was much more relevant to the situation, a couple of female woolly monkeys. He had met and admired Sambo and had confessed to me that he had long wished to try to establish a breeding colony of these delightful beasts at his home in Worcestershire. Not only was he a fanatic about South American monkeys but he had also undertaken a number of expeditions in their home continent.

In addition to all this, he was nearly as keen on birds of prey as I was. He owned something for which my greedy heart yearned and which he was ready to exchange for Sambo. This was a Black Collared Hawk or Fishing Buzzard from the banks of the Orinoco. About the size of a small European

buzzard, 'Bianca' was glamorously coloured for a hawk. Her head was of the rich Cornish-cream hue that one normally associates with the mane and tail of a Palomino; her body, a deep warm mahogany red, ticked with black flecks; whilst across her upper breast she wore the broad black band that gave her her specific name. However, the most remarkable thing about her was her claws, which were enormous black grappling-hooks that seemed designed for a bird twice her size. At first sight these talons did not appear to be particularly sharp and seemed remarkably ill-adapted for grasping and holding on to the agile slippery prey which, according to the books, constituted the bulk of her quarry in the wild. I was soon to discover how deceptive appearances were.

Bianca had obviously been hand-reared and was accordingly tame almost to the point of domestication. She lived loose about the garden, and about the house too if she managed to find an open door or window. She was exceedingly noisy, had a whole series of mewing, clucking and crowing calls and, when particularly interested in something, would bow her head, fan out her rather stubby tail, extend her wings at right angles to her body and shout out something that sounded exactly like 'Chichester!' This was so loud and emphatic that I almost wondered if she had ever lived in that ancient and beautiful city and was suffering from a prolonged attack of home-sickness.

I, personally, took Sambo down to Worcestershire, because I wanted to see what his future home would be like. I was both astonished and delighted. The monkey-house would most certainly have out-pointed most hotels that I have stayed in for pure luxury. It seemed to have everything, including its own swimming pool. The living quarters were built of brick and heated by a series of pipes and electrical gadgets which kept the temperature to almost Amazonian heights, while the outside exercise area, which seemed to cover half the enormous garden, was enclosed by chain-link fencing and included, amongst other items of entertainment, a number of trees and shrubs, a collection of ropes, rubber tyres and trapezes of which any gymnasium would have been proud, whilst a natural stream flowed right through the middle of it; in fact, so luxurious was it that had I been able to ensure a regular supply of *Field* and *Country Life* I wouldn't have minded moving in myself.

The fact that Sambo had two prospective mates was all to the good. Most monkeys are polygamous, given half a chance, and patriarchal as well, and they certainly seem to thrive better in small groups headed by a benign, if dominant, male. For the time being, he was put into a smaller acclimatization cage, so that he could see but could not touch, until he had got the general idea as to what his purpose in life was to be. I helped settle him in to his new quarters and said goodbye to him. He was rather unflatteringly philosophical about my imminent departure, but he had never been a particularly one-man monkey at the best of times. If he had been, I doubt if I could have parted with him, so that this situation would never have arisen.

Bianca was sitting on top of a revolting piece of statuary in the middle of a small pool which formed the centre of a rockery and its collection of alpine flora. I did not explain to her that she had become involved, entirely involuntarily, in the monkey-dealing business, but picked her up rather unceremoniously with the idea of putting her into the travelling-box just vacated by Sambo with as little fuss as possible. Bianca had other ideas. She didn't think much of me and was quite happy fooling about in her pool pretending to catch great-crested newts, and so, just to air her feelings, and to show that she did not intend to be trifled with, she bit a piece clean out of my right thumb, at the same time sinking her enormous talons into the back of my left hand. This was not a promising beginning to our relationship but at least it showed that Bianca had courage and character and had no intention of submitting tamely to any buffetings that fate might have in store for her. As it happened, this was the only time during more than two

99

years that she lived with me that she showed the slightest resentment at being handled.

At the shop I turned her loose in the large sunny room previously occupied by Rusty and the other canine lodgers. I fixed up a perch at each end of the room to give her plenty of flying space and installed a large, shallow, stone sink as a bath and source of drinking water. She proved to be distinctly aquatic in habits and spent much of her time paddling and splashing about in this rather unexciting substitute for the Orinoco or simply standing meditating on a rock in the middle of it, peering into the water in the hope, perhaps, that a piranha, or some other denizen of the tropical deep, would appear and awaken her deeply implanted piscatorial instincts.

In order to give her some interest in life I used to slip a fresh herring or a handful of sprats into the bath. She appeared thoroughly to enjoy the game, pouncing on her long-defunct quarry again and again, sending flurries of water fountaining all over the room, sometimes almost disappearing beneath the surface in her enthusiasm. Finally she would throw the fish ashore, leap on to it, grab it in one foot and hold it up for a thorough inspection before devouring it with an almost infectious enjoyment.

Her tastes were by no means confined to fish and she once made exceedingly short work of a couple of half-grown black rats that had been unwise enough to emerge into her quarters from a hole in the skirting-board. Her speed and agility in this case were such as one might have expected from a cross between a goshawk and a working terrier. She pursued her quarry on foot, assisted by partly open wings, accounting for one of her prey by a spectacular leap as it was rapidly shinning up a curtain that I had left hanging to give the room a suitably occupied look in case unwanted intruders should happen to call in my absence. Incidentally, these black rats, though believed to be almost extinct elsewhere, and probably living up to their old name of 'ship rat', coming ashore from vessels from the continent, were by no means uncommon in this part of London. They were more common in fact than the larger and even more revolting brown rats, which could be seen in hordes, pilfering scraps on the banks of the Thames after darkness had fallen, where they appeared to be entirely indifferent to the movement of traffic and onlookers close by.

I was overjoyed to receive Bianca. Not only was she quite different from anything I had seen before, she represented an entirely new continent. South America has always seemed to me to be the naturalist's Mecca, far more so, I regret to say, than Africa, which, much as I love it, somehow appeared a bit overdone – probably the reason why Conan Doyle preferred the former as a setting for *The Lost World*, which in my opinion has always been the story of stories. I doubt if there are many pterodactyls lurking amongst the Ruppell's Griffon vultures at their rocky columns at Hell's Gate, near Naivasha in Kenya, but in South America I feel that anything could happen. I might add that I have never been there yet.

I now had two birds of prey and life seemed to have little left to offer. Random, on the other hand, took a jaundiced view of Bianca and made it plain that, given the slightest opportunity, she would bring the new relationship to an end with as little delay as possible – but then, Random, throughout the fifteen years that I have been owned by her, has made it flatteringly obvious that she has no intention of tolerating anything or anyone that could possibly come between us.

With Sambo happily settled in the country, I immediately put the business up for sale and began to look for somewhere else to live. I was astonished at the number of people who wanted to own a pet-shop and even more astonished at their motives. Within twenty-four hours of my advertisement appearing in the 'Businesses for sale' column of a well-known evening paper, prospective pet-shop owners began to arrive in droves – and a remarkably unattractive lot they proved to be. It is true that few of them could have had the same reasons for entering the livestock business as I had. Indeed, as a close if somewhat cynical friend of mine put it, 'You must be the only person in the whole history of commerce who bought a pet-shop simply as an excuse to keep pets!' Most of those who presented themselves at my door seemed to be entirely and unromantically interested in the question of profit and loss, a subject to which I regret to say I gave little thought; a fact that was only too apparent from my bank statement.

One particularly obnoxious fellow had the effrontery to suggest that my premises would make a good holding ground for animals, particularly cats and kittens, which would eventually find their way to hospitals and laboratories. Just restraining my temper, I showed him round the property. I opened a certain door at the front of the shop, omitting to mention by whom it was occupied. He was wearing a black coat and a trilby. If there is one thing on this earth that Random detests, it is men wearing black overcoats and trilbies. She doesn't care much for vivisectionists either and in any case objects to being intruded upon without a formal and lengthy introduction. I ushered the man into the room. Somehow the door slammed and the handle fell out. I heard the crash as Random landed, followed by the shuffle of running feet and a sinister silence. I fought to open the door. The handle grudgingly turned in my fumbling hand and I burst into the room. The man was standing rigidly in a corner whilst Random, a foot or two in front of him, was mantling horribly, her head thrown back and her mouth wide open, as, with her golden hackles raised like a war shield, she defied him to move a step. Thankfully I noted that she had been content with this awe-inspiring demonstration and had not carried home her attack. I called her to my fist and opened the door, while Random's foe made good his getaway.

I felt beholden to apologize for this incident and the man left without much ado, black coat and trilby still intact. On his first arrival at the shop he had given me his name and address and I was not altogether surprised to read a few weeks later in one of the local newspapers that he had been charged with stealing a number of·cats from the locality which he had intended to sell to hospitals. In his defence he had had the impertinence to say that he regarded his victims merely as so many twelve-and-sixpences on four legs. He did not mention his encounter with Random and I heard no more of the affair.

Eventually I sold the lease of the shop to a pleasant little man who wanted to sell carpets and had not the slightest intention of becoming involved in the livestock business. His only stipulation was that I should stay on until I had disposed of the remaining stock, both animal and foodstuffs. This arrangement was mutually satisfactory, as it gave me time to find somewhere to live. My grandmother still had her flat in Wilbraham Place but I could hardly install myself there even on a temporary basis accompanied by Tally and two large and exceedingly active raptors.

Meanwhile, I scrutinized the advertisements in every paper that I could lay my hands on. The variety of businesses offered for sale was astounding but none of them seemed to me the sort of thing that would suit my somewhat limited talents. I could not see myself as the proprietor of a string of taxis in Walthamstow; a betting shop in Barking seemed more fun, but I was a bit put off by reading that the owner of one of these establishments had been blown up and half killed by local racketeers because he had valiantly refused to stump up the required sum at the right moment. Finally something caught my eyes that really appealed to me. The advertisement read something like:

Well established Boarding Kennels in Attractive Country District on banks of the Thames near Reading. Genuine Reason for Sale.

This, I thought, was me all over. I could see myself strolling along the tow-path in the evening sunlight, the white may bushes high-garlanded with creamy blossom, with Random on my fist and Tally at heel, whilst assorted hairy lodgers cavorted happily around me, and the pipistrelles pirouetted overhead, returning tired and content to count my takings by the gentle glow of oil-lamps in my caravan. (Two large caravans were included in the deal.) So intrigued was I that I hurried down to Reading to have a look at this paradise.

Everything was, at first sight, much as the advertisement had described it. The Kennels were, in fact, nearer to Shillingford Bridge than Reading, but this was after all of little consequence. The property comprised about three-quarters of an acre, which, though it may not sound much, is surprisingly large when you actually see it. It was enclosed by green palings, backed by diamond-shaped wire netting, and in fact looked rather like an enormous hard tennis court. There were two exceedingly well-equipped and serviceable-looking caravans, which somehow evoked an image of a latter-day Toad of Toad Hall, with their suggestion of airy impermanence. As I stood looking around I could feel the open road calling me.

This, however, was not the general idea of the transaction. There were a number of well-made and comfortable-looking kennels, though it did strike me as rather odd that there appeared to be no occupants. For some reason I omitted to mention the fact at the time. The price asked seemed to me to be reasonable enough and I could see from a glance at the books that there had been a fairly satisfactory number of clients. When I mentioned the reason for sale, the owners told me that they were running an hotel in Reading and that they had found it impossible to combine the two businesses satisfactorily. I told them that I would think the matter over but was a little perturbed when they asked me for a large deposit on the spot. Pondering deeply I returned to London and for once decided to seek legal advice.

Soon the 'Genuine Reason for Sale' came to light. What the would-be vendors had omitted to tell me, and what my solicitors soon established, was that the licence to run these kennels was just about to run out and that according to the local council there wasn't the slightest chance of it being renewed. So that, for the princely sum of £2000 or thereabouts, I would have acquired three-quarters of an acre of rolling Berkshire tow-path, two reasonably well-appointed caravans, a few useless kennels, in which I suppose I could at a pinch have housed my guests, if any, and the prospect of starving to death whilst contemplating some of the loveliest riverside scenery in the south country. I have often wondered how, had I bought the property, the ex-owners would have explained the situation away.

Chapter Ten

Not long after this disappointment, another advertisement appeared in a national daily newspaper:

Small Grocery and General Stores in delightful Old World Fishing Village on the South Coast.

The last thing on this earth that I have ever contemplated was the possibility of entering the grocery business. However, the prospect was not entirely without charm and, I considered, well within my capabilities. Surely, if I could sell hamsters, I would have no great difficulty in disposing of ham? I pictured myself leaning negligently against a rich, deeply-polished mahogany counter, surrounded by a large and admiring circle of fishermen's seductive, if unsophisticated, wives and daughters, their baskets bulging with the products of my well-stocked shelves and their eyes popping with wonder at the stories of my doings in the Dark Continent with which I was regaling them.

This vision appealed to me so strongly that I wrote to the box number supplied. Within a few days I got an answer and arranged an appointment to view the property. The following Sunday I left Waterloo station on my way to keep my new appointment with destiny. To judge by the size of the car that met me at Havant, the owner of the store must have been doing extremely well. (Later I discovered that he had hired it for the occasion and that he owned no car whatsoever.) However, it had the right effect on me. He must, I thought at the time, be selling an awful lot of baked beans and slices of cheddar — and if he could do it, why shouldn't I?

The 'Old World Fishing Village' turned out to be Emsworth, which is certainly attractive enough provided you happen to live in the right part of it. The fishermen have mostly been replaced by week-end yachtsmen, but there is still a splendid hint of nautical adventure in the names of some of the houses along the fore-shore. 'Galleons Reach' and 'Moonrakers' seemed redolent with their suggestion of piracy and smuggling and indeed some of the inhabitants, heavily bearded and sea-booted, might well have descended directly from those who sailed with Blackbeard and Captain Morgan, even if their present occupations seldom took them beyond the confines of Langston Harbour. The place certainly had all the right ingredients. Small craft bobbed and dipped at anchor like resting water-fowl. Herds of mute swans inspected the waving fronds of sea-wrack for delicacies or waylaid admiring visitors in the hope of a hand-out, and on the roof of the Yacht Club perched row upon row of clamorous black-headed and herring gulls, pearly grey and silvery white in the strong sunlight. It was, in fact, about as maritime a place as one could wish for.

However, the shop, the sole object of my quest, was, as I might have guessed, far from this scene of timeless tranquillity. It did not look out on the swelling grey waters of the English Channel, but offered instead an unrivalled vista of the comfortably rounded figure of the local gasometer, which was a magnificent specimen of its kind. The railway station was well within whistle distance to the north, and past my door ran a tributary of the main Portsmouth road. None of this did much to encourage the latent poetry in my soul but, after all, one cannot live on scenery alone. I was here to do business. I had, moreover, noticed that real country, wooded, hilly and enticing, lay within easy cycling distance to the north and east.

The shop itself was a sort of homely affair with its hint of friendly personal service that one sees in so many television advertisements (the fact that the vast majority of the population prefer to deal with the supermarket in the square is no doubt merely a sign of the times). The account books seemed to be reasonably optimistic and showed what I had hoped — that whilst I was unlikely to make a fortune I should at least be able to support my ever-growing family of assorted beasts. The living accommodation was spacious enough for my purpose. There were two large and comfortable rooms upstairs, one of which would make a splendid home for Bianca.

What was still more to the point was that, adjoining the private part of the building, was a long narrow yard (at the time masquerading as a kitchen garden) which led into a large well-lit empty garage, an ideal combination providing a splendid mews and weathering ground for my two birds of prey. The fact that one side of this garden was skirted by the main road and the other by a similar garden belonging to my nearest neighbours did not unduly worry me. How the neighbours would react when they learned that henceforth they would live cheek by jowl with a golden eagle, a South American fish hawk and an enormous and exceedingly agile saluki didn't perturb me particularly either. Doubtless things would work themselves out in the fullness of time, as they had always done before.

I considered that this undertaking would at least be an improvement on the pet-shop business and so, without much persuading by the present owners, I decided to give it a trial, provided that my solicitors did not unearth any horrors of the sort that had been brought to light over the ill-fated kennel proposition in Berkshire. Thus resolved, I returned to London, content at the thought of what I hoped would prove to be a placid if not madly exciting future.

Shortly after my first arrival in Bermondsey, I had, whilst exercising Tally in Southwark Park, met and talked with an ex-paratrooper named Bill Lynes, who was similarly employed with his bitch, Penny, an engaging reddish-brown beast whose ancestry was doubtful but which, from her large erect ears, watchful expression and general demeanour, obviously included an alsatian amongst her immediate forebears. Bill was, at the time I knew him, the landlord of the Prince Imperial in the Rotherhithe New Road and Penny was the self-appointed guardian of that cheerful and hospitable hostelry, a role which suited her admirably. She struck up an immediate friendship with Tally and would rush across the wintry wastes of Southwark Park to greet him during our mutual lunch-hour safaris to that rather dismal piece of *rus in urbe*. She might have been fond of Tally but she never re-garded me with anything but the gravest suspicion and, even after months of close acquaintanceship, would just about acknowledge my existence.

Bill and I had often discussed the possibility of a mating between these two and speculated as to what the resulting progeny would be like. In due course the union took place and exactly nine weeks later Penny gave birth to five huge, hungry and hairy puppies. Four of these were obviously miniature salukis, exactly like their sire must have been at the same age, whilst the fifth, a bitch, was the colour of autumn leaves, a rich golden coppery hue. Even then, her long legs, pointed muzzle and slender tail showed that still the potent Eastern greyhound blood was coursing through her veins. It was this puppy that I chose and I named her Bracken on the spot, partly because of her colour but more in memory of the lurcher bitch, my first dog, who had been with me for so many of my early years and who, by her very presence, had helped to keep me going when life was at its bleakest.

A day or two after my visit to Emsworth Bill telephoned to tell me that Penny's family were now more than six weeks old and were becoming increasingly lively and obstreperous. He would be pleased if I would collect Bracken, who, although a female, was the biggest and sturdiest of the lot. Penny, it seemed, was beginning to show signs of strain, had nearly run out of milk and was making it increasingly obvious that the joys of motherhood were wearing thin. I went over at once and, although being nearly savaged by Penny, who it was quite obvious trusted me even less than usual, I stooped amongst the litter and picked up Bracken, who was nearly as broad as she was long, and tucked her carefully into a new zipper bag that I had bought especially for the purpose.

'Whilst you're here,' said Bill's wife, 'I thought that perhaps you might be interested in this.'

She led me to their large comfortable sitting-room at the back of the pub, where a huge fire was glowing, and pointed to an enormous arm-chair. Curled up, nose tucked between forepaws, was one of the smallest dogs I had ever seen. Battleship grey, with a snow-white chest and a broad collar of the same hue running half way round the neck, with huge pricked ears and protruding eyes like those of a bush baby, it took me back to the time when, in Africa, I had reared an Oribi faun called Swara. On seeing me the little dog stood up, stretched luxuriously, jumped to the floor and trotted up to me on tiny slender legs that should by rights have ended in hooves instead of paws. He was an Italian greyhound, just one year old, and his name, as one might have guessed, was Bambi.

Molly Lynes had noticed the little dog sitting day after day in an upstairs window of a shabby house close to the Prince Imperial. She had seen the owner passing several times with a brand-new poodle puppy, complete with rhinestone collar, trotting at the end of a smart new lead. From all this she had drawn her own conclusions. One evening she had knocked at the door and made inquiries. The woman who appeared from within told her briefly that she was tired of the little grey dog, had in fact never really cared for him, and in any case had always wanted a poodle (a breed which at that time was riding the crest of the canine popularity wave). Molly, who could be a formidable proposition when she wished, begged or bullied her into handing the dog over there and then and took him home to the pub, where Penny promptly tried to murder him before giving way to a fit of sulks and refusing to eat a morsel. Now, having primed me with a few glasses of whisky at Bill's expense, Molly without warning asked me if I would bear the

Italian greyhound home, love and cherish him and take him into my heart.

Molly's nostrils at this juncture were flared like those of a war-horse scenting battle and I knew better than to refuse, besides which I was intrigued with the little blighter, who was utterly unlike anything that I had ever encountered. Bambi in the meantime had clambered on to my knee and had thence struggled bodily under my pullover and was now pressed snug and warm upon my chest. I hesitated, but before I could say a word a new lead had been produced and I found myself out in the street with Bracken mewing in her bag and Bambi prancing along beside me, worrying the lead and dancing with the characteristic high-stepping action which is peculiar to the breed and is shared only by the well-bred hackney.

Whatever Bambi lacked in size he more than made up for in impertinence. Tally inspected him all over in the aloof, impersonal saluki way, whilst Bambi, who was famished, bolted more than half his dinner, an act of *lèse majesté* that the larger dog chose to ignore. The frail little hound, who looked as if he would be unable to withstand the slightest gust of wind, was utterly fearless. He loved Tally from the start of their acquaintanceship and Tally, for his part, allowed him to take liberties which he would have punished instantly if perpetrated by any other dog.

Random, on the other hand, took a different and quite basically gastronomic view of what she obviously regarded as a curiously shaped and under-sized antelope kid. Once in my stupidity I let Bambi out into the yard, forgetting that Random was already in occupation. Within seconds a muffled scream brought me skidding and stumbling on the scene. Random had left her throne in the sunniest corner of the yard and had grabbed the unfortunate little dog by one hind leg. When I arrived she had completely immobilized him. Instinctively carrying out the only effective action, I seized a bucket of water that by good fortune was standing in a corner and flung it all over the indignant eagle. She dropped her prey and went for me whilst Bambi scuttled three-legged through the door to safety.

Having eaten until he bulged, Bambi took careful stock of his new surroundings before scrambling up the old rickety stair-case and disappearing. An hour later I found him curled deep inside my bed, his little body glowing like a hot brick. For the next ten years or more, my bed was his bed, to our mutual advantage. Italian greyhounds make splendid hot-water-bottles, their normal body temperature appearing to be several degrees higher than that of any other dog, and they have one great advantage in that they do not become cold at 2 a.m. and have to be refilled. Once when I had an unusually vicious attack of flu he slept pressed close to my aching body for over a week, leaving only to answer the calls of nature, slipping out into the garden by his own special route without even waking me.

After the usual boring but necessary legal negotiations, the Emsworth shop, fixtures, fittings and stock – the lot – became mine, to do with as I wished. In Bermondsey, the cleaning-up operation and the ceremony of handing over the keys took time, but at last I found myself in a large pantechnicon borrowed for the occasion together with its driver, heading south-westwards with hope in my heart and precious little in my pocket. In the back alongside my few possessions, which seemed to consist almost entirely of books of all sorts which had been with me for years, had travelled widely and showed the results of it, Random and Bianca were concealed in their boxes, whilst Bracken, blissfully unaware of all that was going on around her, was curled fatly amongst the tattered pull-overs in her special basket with its wire grille front. Bambi balanced on my knee, shouting abuse at any dog that he considered was within insulting distance, whilst Tally, dignified as ever, peered over my shoulder, blowing great warm breaths down my neck, to show that he was very much part of the expedition.

My chauffeur was a cheery cockney whom I had met several times and whose dog, Spike, I had looked after for several short periods. Reticence is not one of the true cockney's most noticeable characteristics; none the less, I was surprised, to put it mildly, when, somewhere between Hammersmith and Putney, he announced casually that he had recently been enjoying Her Majesty's hospitality – or, as he put it, 'doing a bit of bird' – for selling pornographic literature. For one who collects unusual characters as others collect postage stamps here indeed was a prize specimen, and he kept me enthralled with his experiences and his views of the British penal system for the next two hours or thereabouts, until, in fact, we pulled up with as much of a flourish as the aged and rather dilapidated vehicle could manage outside the door of my new emporium. Within a few minutes we had unloaded all my possessions, such as they were. The late owners were still there to hand over officially, and a fire, to me at that moment the most important thing of all, was flickering in the grate of what was to be my chief sanctum and refuge from the cares of high finance.

Tally and Bambi, stiff, tired and thirsty after their journey, had to be taken for a short walk down a lane that might have been made just for our purpose. Starting close to the pub on the opposite side of the road, it ran between fields in which herds of Friesians and Shorthorns were grazing and a thick, tangled hawthorn jungle, which incidentally struck me as being an ideal site for a hawfinch's nest. Winding its way onwards, dominated by the ever-present bulk of the gasworks, it came to a foot-bridge across the Em, the little swift-flowing stream that gave the town its name.

Random, delighted to leave her box, was soon sitting on her block in the garden, her great wings opened to the last rays of the evening sunshine, her head craning to the full stretch of her neck as she took in her new surroundings and weighed up the possibilities. Bianca, after a therapeutic bath in the kitchen sink, was soon happily demolishing a fresh herring in the room which I had allocated to her as her special mews home. Harry, the driver, had served for several years in the Royal Navy and had not come unprepared. From somewhere deep in the interior of his enormous greatcoat he produced half a bottle of rum, which we rapidly disposed of.

There is to me something particularly satisfactory about having a celebratory drink with a friend or, as in this case, a congenial acquaintance, knowing full well that the friend or acquaintance will, in half an hour or so, be well on his way, leaving me to consolidate my thoughts and re-arrange my emotions in pleasant solitude. Harry and I drank our rum, rich, warm and spicy, and swore everlasting friendship, although we were both perfectly aware, even as we spoke, that we were unlikely ever to meet again. Harry was undoubtedly a rogue but he was also a character, and there are none too many of those about these days. I watched the great pantechnicon lurching and swaying like a mechanized mastodon as it swung round the corner and out of sight. My final link with London had snapped and a new era was about to open.

I returned to the fire where Bambi and Tally were snoozing happily and twitching in their sleep. Like Kipling's cat that walked by itself, all places were alike to them, provided I was there to share it. There was about half an inch of rum left in the bottle, which was standing looking rather forlorn on the mantelpiece. I drank it with a curious blend of contentment and speculation. I was finally committed to my new way of life and there was nothing I could do at this juncture which would alter the situation. I settled into my old, battered arm-chair, one of the few pieces of furniture I had brought with me, only to jump up again after a few moments to light the gas in order to heat a saucepan of milk for Bracken's evening meal. She was then getting four meals a day and growing almost as I watched her. With a family such as mine to care for I had little time for reflection.

I heard a gentle, almost timid knock on my door (for some reason the premises didn't boast a bell of any sort). On opening it, I found my next door neighbour, Mrs Feathers, standing there flanked by her two daughters, each holding a bunch of flowers. 'We heard you were arriving today and we wondered if we could do anything to help,' said Mrs Feathers, who as it turned out, was as true a cockney as any that I had left behind in Bermondsey. I asked them to come in and I introduced them in turn to Tally, Bracken, Bambi and Bianca, and finally, as a grand finale, to Random herself. The fact that they were now living, willy-nilly, next to what must have appeared to them at the time as a combined invasion from Crufts and the London Zoo troubled them not one whit. The Feathers family became then and remained amongst the best friends that I made during my sojourn amongst that closely knit community.

Although kack-handed and basically incompetent, it took

me remarkably little time to master the intricacies of a bacon-slicing machine, which despite its humble role in life is none the less a pretty lethal weapon with much the same potential for destruction as a miniature guillotine. It was with a glow of artistic satisfaction that I found myself slicing off paper-thin slices of ham after a short probationary period. I was more than relieved to find that I had a part-time assistant, a splendid if rather unambitious old boy who was then over sixty-eight years old and who, it appeared, had spent his entire working career, except for four years with the Royal Artillery during the First World War, as an employee in a succession of grocers' shops. What he didn't know about the business just wasn't worth learning. After a few days under his patient tuition I felt convinced that Messrs. Fortnum and Mason had nothing on me and, much more to the point, money seemed to be flowing into the till, which was, after all, the whole point of the exercise.

Chapter Eleven

Hidden under a pile of old sacks in the garage, which had been transformed into Random's sleeping quarters, I unearthed an ancient bicycle – a bicycle that I am sure must have superseded the penny-farthing, but only just. It was a wonderful old machine, heavy with years and bursting with character. Had anyone thought of making a film about veteran bicycles on the lines of *Genevieve*, this one might well have rated stardom and immortality. Someone must have cherished it throughout its long career as, apart from flat tyres and handlebars festooned with the dusty cobwebs of years, it was in perfect working order. It was so high that one almost needed a mounting-block to reach the saddle, but what really appealed to me was that there was a great square metal frame suspended over the front wheel. This was obviously intended to carry a grocer's delivery basket, but all I saw at the time was that here before me was the ideal mobile eagle-carrier. With Random perched upon the framework, well padded, of course, against the chilly metal, and the dogs cantering beside me – this was my long-awaited key to the freedom of the countryside.

I lost no time in turning this into reality. By tightly binding a length of strong leather round the framework, I soon had what would pass as quite a respectable self-propelled cadge (a cadge being, in the esoteric language of falconry, a sort of rectangular perch slung from the shoulder of a humble member of a hawking party and used to convey falcons to and from the field and to provide a place on which they can rest, hooded, when not being flown at quarry). Random never objected to being hooded and now this easy acceptance was to prove its worth.

Both Tally and Bambi took readily to this form of exercise and, indeed, Tally quickly learned to regard it as second only to the fun of following a horse. At first we did not go far. On Sunday mornings I would creep shivering from my bed as soon as the first faint-hearted glimmer of grey autumnal light appeared at my window. I would make a mug of tea, which I would carry back to bed to share with Bambi, who would struggle up from his cocoon like a surfacing porpoise. He loved his bed as much as I did but he loved his morning saucer of tea even more. Tally, curled amongst the blankets before the dying fire, would stretch, yawn, and stretch again, and have a good satisfying scratch before slowly rising with much the actions of a camel until finally standing up with his paws on my chest, when I slipped his collar over his head and snapped the lead on. Bracken was still too small to take an active part in the proceedings and would remain at home, happily wrecking the room and reducing any wood she could reach to splinters.

Taking Random's hood from the mantelpiece where I kept it when not in use and where it made an unusual and much admired ornament I would wheel the old bike out of the lean-to next to the kitchen and into the silent, sleeping street. Bambi would then do his best to ensure that it slept no longer. His one exasperating fault was his habit of letting rip with his high staccato machine-gun voice whenever he was excited, which, unless he happened to be asleep, seemed to be most of the time. He was, in fact, as voluble and emotional as the human inhabitants of the country where his breed is supposed to have originated. Luckily I was generally able to stop him in mid-aria by a succession of oaths and endearments, though I must confess there were times when I wished I had settled for a basenji, the African voiceless hunting dog.

As I unlocked the garage door I would hear Random stirring on her perch and the dry crunch as the foot she had been resting tucked inside her lower breast-feathers grasped the branch on which she sat. In the chilly dusk I would feel

for her head and stroke her nape before, with a gentle pressure
behind her legs, I persuaded her to step on to my fist, which
she could feel but not see. I would slip the hood quickly
over her beak, ease it into place and tighten the braces, before
transferring her to the comfortably padded perch of the erstwhile tradesman's basket-carrier. With her leash and those
of the two dogs held in my left hand I would pedal slowly
away into the rapidly dispersing gloom.

A mile or so to the west was a marshy, tree-dotted waste-
land, where sea and country met, divided only by a grey stone
wall, which helped to save the low-lying meadows from flooding when the channel combers rose in wind-whipped fury.
Behind the sea wall were tussocky fields, with shallow pools,
from which a startled spring of teal might suddenly shoot
upwards, moving at a speed which might well defy the fastest
falcon to get on terms with them. Seawards lay the weed-
scattered strand, which when the tide was out revealed
hundreds of mysterious pools in which lurked all manner of
temporarily marooned salt-water life – blennies, gobies and
the curious pipe-fish, looking rather like a straightened-out and
attenuated sea-horse. Picking their way amongst the piles of
glistening bladder-weed were bands of curlews, the loveliest of
all the wading-birds, which would rise upon my approach,
breaking into their wild chorus of fluting, bubbling calls, the
spirit of the moorland and the marshes. Here and there, looking almost too exotic for the grey-green British littoral, could
be seen scattered groups of shelduck, with their harlequin
pattern of black, white and brilliant chestnut patches.

Rabbits crouched amongst the tussocks and would streak
for cover in the almost impenetrable surrounding hedgerow,
getting into top gear from a standing start and defying the
efforts of the far-ranging saluki to get on terms. Hares, not
rabbits, are the natural quarry of these desert greyhounds.
Bambi acted the part of a working terrier with the greatest
enthusiasm, diving into the prickly fastness with complete
disregard for the punishment inflicted on his tender skin by

the battery of thorns and clutching briars. His high-pitched
voice would ring out from the depths of the thicket as he
strove valiantly to push the elusive quarry out into the open,
where Tally was waiting to receive them. As the rabbits knew
exactly what they were doing and where they were going about
their own private runways, casualties amongst them were few
indeed, but the dogs never gave up hope.

Once a hare, a wanderer perhaps from the cornfields to the
north, burst out from the hollow where it had been sheltering
and set off across the rushy fields with Bambi shrieking behind
him. Bambi was certainly no heavier than his intended victim

and indeed he looked very much like it, with his long legs and his great trumpet-like ears cocked forward in the excitement of the chase. For his size he was fast and his stamina was surprising but the hare had the legs of him, and by the time Tally, who had been following a line of his own, came crashing on to the scene, the hare had slipped under a culvert and disappeared into a field of kale, which swallowed him up as completely as if he had dived beneath the surface of some great wind-whipped lake. Bambi was bursting with pride at having routed so large a foe. He had proved beyond doubt that though so tiny he was still very much a greyhound at heart.

After the dogs had had their exercise and had returned panting and smiling-eyed from the farthest points of the marshes, I would leash them up and let them sprawl contentedly in a sheltered hollow beneath the battered sea wall whilst Random had her turn. There was an ancient tree stump, black and rotting with time and weather, which rose like a solitary rock dark against the grey-green background of the marsh vegetation. Random, unleashed, would leave my raised arm and with two deliberate flaps and a short glide would reach this isolated projection. She would grasp it with her golden feet, rouse her feathers and swing round to watch my movements and see what was afoot. I would show her the lure and she would stand on tip-toe, her head craning to see what I would do next.

I would walk the length of the field, holding the lure tucked beneath my armpit. Random's yelping calls would follow me, becoming more querulous and insistent the further I went. I would wave the lure and call her into the wind, and she would be there, poised above my head; her penetrating amber eyes peering downwards into mine; arriving, so it seemed, almost before she had left her launching-post. Should I keep the lure concealed, she would put an aerial girdle round the field, looming huge in the misty morning light, before taking stand on the top of some distant stunted thorn tree and shouting her displeasure at my lack of co-operation. Sometimes I would feint a throw. She would launch off at the first hint of movement from my arm, coming straight in in level flight, challenging me to keep her waiting longer.

Once, to my embarrassment and her uncontrolled rage, the meat fell off the lure and disappeared into the long grass before she arrived to clasp nothing but the tough, ungarnished leather. Her hackles rose and she glared into my eyes with a look of supreme contempt, before chasing me for several yards until I tripped and fell full length into a furrow. Thereupon she walked round me, apparently to ensure that I was still alive, before jumping on to my back and tweaking my ear with her immensely powerful grey-blue beak, chirruping with her gentle conversational tone, satisfied, it seemed, that I had learned a lesson which I was unlikely to forget.

One morning, instead of alighting on her stump, as expected, Random overshot the mark and shot high into the air, higher than I had seen her mount before. A stiffish breeze was blowing from the land but that worried neither of us. She could ride any normal wind with supreme ease and grace. The tide was at its highest and the only sea birds in sight were scattered bevies of black-headed gulls and the occasional herring-gull, drifting by like pieces of wind-tossed paper. High overhead was something neither Random nor I had noticed – a huge bird, six foot across the open wings and with a beak like a pick-axe, was cruising in circles a thousand or more feet up and watching all below.

I had brought that day a special lure of freshly killed wood-pigeon, picked up by one of my customers on the road outside Havant. (This picking up of roadside casualties was a trait I strongly encouraged and a surprisingly wide variety of victims was brought to me as a result.) Random enjoyed pigeon meat better than most things, and the soft loose plumage that can be trying to a hawk bothered her not at all. She took stand on the crown of the tallest oak tree within sight, her dark body, with its lighter shadings, blending at once with the late autumn foliage. Had I not seen her alight I might have walked right past the tree in which she sat. Then, seeing the pigeon, she crashed out of the tree, snapping off some of the smaller twigs as she came. A sudden cross-wind caught her unawares and spun her round, lifting her up as if she was a huge brown paper bag, whirling her suddenly seawards, whilst high above the great black-backed gull, the pirate of the skies, was leaning on the wind, watching her every movement from ivory-coloured eyes.

Random, out of training after her period of partial inactivity in London, was now swinging in a half circle a hundred yards or more over the sea and only ten or fifteen feet above what must have seemed to her a strangely undulating ploughed field. I rushed to the wall in alarm, thinking that she might try to settle upon the heaving water and become engulfed beneath the waves, which seemed to be clutching at her as she laboured to reach the land. The great black-back, ever ready to turn such a situation to his own advantage, circled lower. This gull is a ruthless killer of young or injured sea-birds. Many a storm-driven guillemot and shear-water have been hammered to death by its brutal yellow-hooked bill, flaunting its blood-red insignia to mark the assassin's trade.

Random, fighting her way above the waves, could not get the air space she needed to give her heavy body sufficient uplift, and the great gull knew this. He knew too that if he forced her into the water there would be no need for a *coup de grâce*. He, who had probably been hatched some years before on an isolated wave-buffeted stack off the west coast of Wales, could never have seen an eagle before. All he saw now was a huge brown bird in deadly trouble, apparently helpless, and he meant to turn her into carrion with as little delay as possible. When the waves had done their work he, unscathed, would feast upon her sodden body. He checked, half closing his great angular wings, and stooped like a clumsy falcon straight for Random's back.

Although neither the huge gull nor I, watching helplessly from the sea wall, realized it, Random had never for a moment

123

lost her self-possession. She knew where she was going and she meant to get there. She had seen the gull but had given no indication of the fact. What was a great hulking corpse-eater, this flying hyena, to her, anyway? Not worthy even of her consideration. Her target was the shore-line, only a few wing beats away. She heard the singing hiss of the gull's descent as he drove in for her shoulders. So quickly that the action was over before I had taken it in, she flung herself on her back. One huge sickle-taloned foot shot out, as eagle and gull hit the surface of the sea together. It was a glancing blow and Random's grip failed to penetrate the black-back's greasy and densely-feathered breast. With a twisting wrench the gull hurled himself sideways, leaving a blood-stained bouquet of tiny feathers in Random's clutch. Half swimming, half flying, he dragged himself clear of his intended victim and, skimming the surface, just managed at length to heave himself awkwardly into the air. Slowly gaining height, he disappeared into the murky distance, seeking, no doubt, some secluded promontory where he could rest, re-arrange his dishevelled plumage and nurse his wounded breast.

After striking her attacker, Random had, with one incredibly controlled movement, rolled on to an even keel, just as the first wave seemed to reach out and pull her down. She ditched like a damaged bomber. For a few seconds she rested on the surface, her beak wide open with exertion, her powerful wings fully extended, keeping her afloat like a pair of life-belts. She was but a few yards from safety when I shook off my wellington boots and plunged into the heaving water, barely noticing the numbing chill. To my intense relief I felt the soft spongy surface of a mud-bank beneath my feet as I forced myself forward, expecting at any second to flounder off this natural shelf into the deeper water beyond.

Seeing me, Random shook off the lethargy born of shock that seemed to have overtaken her and struck out towards me, her wings acting as paddles in her desperate fight to reach safety. Somehow we came together and without thought of the consequences I managed to hoist her on to my unprotected left arm, which she gripped with a desperation that resulted in wounds which are clearly visible to this day. I turned and somehow kept my balance, despite the weight of Random and the pain, which must have been much as if my arm had been gripped in the teeth of a bear trap. At length we reached high water mark, where I practically hurled Random on to a piece of timber before washing my throbbing arm in the antiseptic sea water which I had so thankfully left behind. Random was a depressing sight as she perched in a semi-coma on that sea-borne log. Her wings and tail were reduced to mere quills, her breast a tangled mass like sodden sphagnum moss. Only her dignity remained, as usual impervious to all that she had just experienced.

The two dogs, still tethered to the old bicycle, knowing well that something was amiss, were straining at their leads, obviously disappointed at having missed all the excitement. Random, having shaken herself repeatedly like a wildfowler's retriever, was soon feeling more cheerful, and it was with untold relief at the lucky ending to such an unlikely adventure that I was soon pedalling homewards, travelling by alleyways and footpaths to avoid the inquisitive eyes of anyone who might be abroad. Seldom has a hot bath, followed by a bubbling plate of locally-made sausages, been more welcome. In an hour or two Random was as spruce and immaculate as if she had never left her perch in the garden, but she took care never to wander out over the open sea again.

Chapter Twelve

As can be imagined, the presence of a golden eagle in a smallish community soon began to cause considerable interest. Small boys could be seen clinging to the fence behind which she perched in all her splendour. Some, bolder or more ornithologically minded than others, would knock on the door or stop me in the street to put the oft-repeated request – 'Please, mister, can we see the eagle?' Or mothers would come into the shop, where they might or might not buy something, and after a certain amount of hesitation would inquire – 'My children are very fond of birds. Would it be possible for them and their friends to come round after closing time to see the eagle?' This sort of thing was harmless enough, even if it happened a bit too often, but there was also the type who would stop me in the street and remark with painful lack of zoological knowledge – 'I hope that bloody eagle of yours doesn't fly off with my kid!'

Random's first real bid for local fame came when, launched from my fist for her usual evening circuit of the surrounding countryside, she, no doubt regarding the local gasometer as the nearest replica of a chunk of the Pyrenees which she had seen since she left her eyrie, sailed up and landed on the top rail of that unlovely edifice. Having decided that this provided the most suitable look-out post for possible quarry, she flatly refused to come down, despite my pleas, curses and the production of some of the choicest lures I could lay my hands on. As night was falling fast and as Random seemed prepared to spend the rest of her life dreaming her secret dreams atop this man-made Mont Blanc, it was obvious that something would have to be done about it.

First of all, I introduced myself to the custodian of the gas-works, where he lurked in almost Stygian darkness in a cave-like office at the foot of the gasometer. He proved to be a most co-operative fellow. He was very proud of his charge, which he regarded with the sort of paternal affection that country gardeners normally reserve for their prize vegetable marrow. To my great relief he turned down my suggestion that I should climb the ladder that seemed to rise interminably and hor-ribly upwards until lost to sight in the misty darkness far above. He told me that he would have to seek the advice of his headquarters, which happened to be in Portsmouth.

Before making up his mind to telephone his superiors he kept me in a cheery frame of mind by telling me in great detail of the awful things that could happen to the unwary and inexperienced who dared to venture into the unknown world above. If you weren't careful, he told me, you could be sucked into the cylinder through apertures in the sides, never to re-appear. The whole thing was crammed full of seagulls that had been unwise enough to perch where Random was now sitting, unaware of the ghastly fate that could be hers. (I still haven't the faintest idea if there was any truth in what he told me.)

The authorities in Portsmouth advised us to contact the police, who arrived in the person of our local bobby. He was amused by the whole affair and inclined to treat it with un-timely flippancy – but then, after all, it wasn't his eagle sitting up aloft in danger of instant engulfment. Having established that there was no law prohibiting eagles from perching on gasometers, and after further humorous asides, he pedalled off to contact the RSPCA.

The representative of that splendid society arrived at length with what, to my horror, appeared to be a wickerwork cat-basket. However, it was not for me to question how he proposed to cram a golden eagle, measuring over three feet in length and weighing more than twelve pounds, into a basket which seemed to me about eighteen inches square. He was the expert — and no mere inspector, either, but a chief superintendent or something equally august. Meanwhile, the Southern Gas Board, showing an unexpected measure of understanding and sympathy, rang back to say that the gas cylinder should be lowered to its full extent. Lowered it was, slowly and imperceptibly, with Random, statuesque as ever, still perched majestically aloft, indifferent to the floodlighting which now bathed the scene in dramatic brilliance and made the ensemble look like an impromptu display of *son et lumière*.

Sensing the unusual activity and the undercurrent of excitement, the local citizenry began to gather in ever increasing numbers. On discovering what the fuss was all about they began to offer facetious pieces of advice, such as — 'Would you like to borrow my little boy's butterfly net, mate?' or 'How about hiring a helicopter from Thorney Island to pick her off the railings?' A representative of the local newspaper must also have been lurking there unseen by me, because the next issue carried headlines that ran — 'Emsworth eagle immobilizes gas-works' — or words to that effect, followed by a sensational and inaccurate account of Random's activities.

The cause of all this entertainment, possibly deciding that publicity was all very well but that things were beginning to get out of hand, without the slightest warning launched herself into the air, looming like an enormous bat in the strange glow of the artificial lighting. She circled once over the heads of the spectators before turning and disappearing into the shadows of the night, which swallowed her up as completely as if she had never existed. I heard a rustling crash as she settled awkwardly in the top of one of the trees that lined the banks of the little River Em. I knew that she would be safe for the night and was unlikely to get into any further trouble for the time being. That, for the moment, was that. Indicating to the lingering crowd that that particular performance at least was over, I thanked the gas-works man, the RSPCA representative, the police and anyone else in sight who had been in any way involved with Random's disgraceful truancy.

I slept my usual untroubled sleep that night, knowing full well that Random and I would meet at dawn, that she would come down to me, and that, happy in our reunion, we would wend our leisurely way homeward through the dripping grass, fending off the groping branches that sought to bar our way. So confident was I that I failed to wake as I had intended and the light was well advanced when I finally collected my glove, lure, hood and all the other paraphernalia. Having assembled my gear, I made a cup of tea and sauntered into the garden to see what the weather was like. I bent to examine Random's block to ensure that all was ready for her home-coming. As I straightened up I heard behind me the creak and whoof of powerful wings. Turning my head, I was just in time to see her swing in and pitch on the garden fence, where she sat talking to me in her own low and surprisingly musical tones. I untied the meat from the lure and gave it to her piece by piece from my fingers. She was not unduly hungry and took the food gently without a trace of her usual somewhat demanding manner. I slipped the leash through the slits in her jesses and lifted her off the wall. She bounded on to the block and began to preen happily. Strangely enough I was not particularly surprised to see her return voluntarily to base. I had known for a long time that she was fond of me and I rather suspected that she regarded the confines of the small and unromantic back garden as her territory, to be looked upon with proprietary pride. There was still a great deal I didn't know, but I was beginning to learn a lot about eagles in general, and about Random in particular.

Hardly had I finished breakfast when the telephone began to ring. People had seen the headlines in the local paper and were anxious to know what the outcome of Random's adventure had been. Had she been retrieved, and if not, what did I propose to do about it? Other newspaper reporters rang up, keen to get pictures of us together, and finally Southern Television got through to inquire whether Random and I would be prepared to travel to Southampton that evening for an interview. I had appeared on television before, about twenty years previously, in a programme entitled *Picture Page*, with Joan Gilbert, Leslie Mitchell and Cressida, the kestrel, shortly after our return from our sojourn in the German POW Camp. This had been followed by a further appearance in an early quiz show with Lionel Hale and once again the photogenic, much-travelled and imperturbable Cressida. In those days there must have been a viewing audience of a few thousand at the most and the screens were about the size of postage stamps – or so it seems in retrospect. Now things were very different, besides which, I had no idea how Random would behave in a studio.

However, I agreed to appear and we duly set off in an immense limousine, hired for the purpose from a local garage, the proprietor of which treated us with the deference mixed with awe to which he no doubt considered potential television personalities were entitled. At the last moment, thinking that Southern Television might as well have their money's worth, I decided to take Bianca along as well. After all, she was in all probability the only representative of her species outside the American continent.

On our arrival at the studios in Southampton we were met by an enthusiastic reception committee. In those days, so-called wild-life programmes were not by any means as common-place as they are today. We were shown into a lift, the two birds each concealed in their own travelling-boxes, with the care that might have been reserved for the Crown jewels. I was ushered into a dressing-room and left to my own devices. At this juncture I was beginning to suffer from an acute attack of stage fright. My heart leapt horribly when I heard a knock at the door. Opening it, I saw gathered in the passage what looked like an army of technicians, fellow performers, make-up girls and all sorts of men who might have been any-thing from film directors to canteen managers. They had collected together for one purpose – they wanted to see the birds.

Random was already perched on the back of a chair, making faces at herself in the mirror. She responded to her visitors with all the graciousness of a monarch receiving a deputation of ministers. Bianca was retrieved from the depths of her box and likewise made an immediate impact on her beholders. With her rich chestnut body plumage, creamy head, and the striking black belt across her chest, her inky black beak and feet with enormous talons, she was in her way as attractive and even more unusual than Random. Had colour television been invented in those days I feel sure that she would have been an instant success.

After wallowing in all this reflected glory I was beginning to feel better, ready in fact for anything, a state of mind which a visit to the TV Club did much to reinforce. My interviewer, Jonathan Miller, and I ran through a short rehearsal of what was to take place before the cameras. It was to be the usual sort of question and answer affair, but it was decided, as a finale and to introduce a spice of excitement and originality, that just before our part of the programme ended Miller was to ask me if Random would fly the length of the set. A strong, heavy chair was brought on and placed at a suitable distance from where we were sitting, bathed in the brilliant and rather uncomfortably hot glow of the lamps. Twice I raised my arm and twice Random spread her great sails and floated on to the chair, as if she had been doing this sort of thing all her life. Everything seemed set for the actual performance.

When the theme music of *Day by Day* ended, Miller and I were 'discovered' sitting nonchalantly in chairs behind a large table. I was holding Random, who panted a bit and looked as if she wasn't too happy about her part in the performance. Miller held Bianca, who seemed quite at home, revelling in the heat, which possibly reminded her of her home on the banks of the Orinoco. I was asked the usual questions – 'How old is Random?' 'Where did she come from?' 'What is her wing-span?' – followed by comments on her recent activities. There followed a few questions about

Bianca, who seemed to be enjoying herself methodically destroying the sleeve of Miller's Fair-Isle sweater.

Came the final remarks. 'She is certainly a splendid specimen of a golden eagle. Could we see her fly?' 'Certainly,' said I, 'Watch this!' and cast her off in the general direction of the improvised landing post. She certainly flew all right. She shot over the chair straight for the cameras, whose operators dived out of the way like startled stags. Zooming higher and higher, ranging in narrow circles, not in the least put out by being enclosed by four walls and a roof, she finally came to rest on the highest point she could reach – a pinnacle of masonry close to the great domed roof of the building. The cameramen had

returned to their cameras, which now picked her out, a distant blob almost unidentifiable against the dusky background.

The interviewer had other subjects to deal with and as such programmes have to be run on a tight schedule I was left to cope with the situation to the best of my ability. I felt a deep resentment against the machinations of fate. To get landed in this sort of predicament twice within twenty-four hours was, I felt, a bit thick. However, something had to be done. It would have been useless to have shown Random any sort of lure. She would have been unable to see it from where she now perched, wrapped in shadows. Besides, after the meal she had eaten earlier in the day, she would not have been hungry enough to be tempted into the hazardous descent.

A horrible-looking fire-escape ladder climbed to within a foot or two of where Random sat aloof. This in itself was a remarkable enough coincidence, because there were several other equally tempting perches at about the same height on which she would have been virtually inaccessible. Thus it was that I, who could hardly stand on the seat of a chair without suffering the severest pangs of vertigo, donned my glove and, breathing silent curses on the motionless figure above, prepared to dice with what seemed to me at the time to be almost certain death.

It was not quite as bad as it had first appeared. The iron ladder was rock-steady. The rungs were mercifully close together and there was a hand-rail which I grabbed frantically at every step. Random in fact was not as high up as I had thought. Still, she was quite high enough up for me. Remembering the little that I had read about mountaineering I gazed steadfastly upwards, trying to ignore the sweat that saturated the palms of my hands. She was looking down at me, intrigued and apparently amused at my laborious attempts to reach an altitude which she had gained in a few seconds.

Eventually I stood within arms' reach of her and began to talk quietly and persuasively. I didn't want to upset her in any way, in case she decided to seek a more secluded shelter and

134

the whole painful process would have to begin again, only much, much worse. As I talked, she began to answer me in her own low, musical, piping voice. At least, she seemed glad to see me. I had to persuade her to stay where she was until I could reach out my gloved fist and ease her aboard it. As I kept on chatting to her I felt like one of those courageous characters who talk down would-be suicides from the roofs of twenty-storey buildings. However, there was nothing re- motely heroic about me; I wasn't enjoying myself one bit and was aching to feel once more the solid ground so far below.

Clinging to the ladder with my right hand I stretched out my left and, pushing it gently against her feet, persuaded her to step on to it as she had done so often before. Even then, I was afraid she might over-balance, due to some clumsy move- ment of mine, and be forced to take to the air again. Somehow I caught the jesses between my fingers and, holding my arm with Random on it close to my chest, began the slow, laborious descent. Had she bated, as she might well have done, I could have lost my balance and crashed down to the floor, still thirty or more feet below. I could visualize the headlines next morning – 'Eagle man hurtles to death in television studio' – followed by all the lurid details of my last few seconds of life. However, she didn't bate. She remained rock-steady and at length the nightmare business came to an end. Never has terra firma felt so solidly reassuring.

I returned to the set to collect Bianca, who was still sitting on a chair just out of camera range and looking disgustingly smug and self-satisfied. She turned her head, first on one side and then the other, looking at me and then at Random. 'Trust a great ugly brute like that to let you down – and in public too!' My chauffeur was waiting for me and together we carried the boxes to the car, pausing only to collect my fee from the receptionist on the way out.

Chapter Thirteen

Affairs at the shop now returned to an even tenor. I was beginning to get the hang of things and to wonder why I had not thought of grocering before. I saw myself as a legendary figure – the Grocer Naturalist of Emsworth; a sort of latter-day Gilbert White. I could picture myself in the years that lay ahead, white-bearded, venerable and benign, trundling my equally venerable bicycle, its basket crammed with packets of frozen food, along the leafy Hampshire lanes, my eyes ever alert for the coppery, flicker-tongued form of April's early slow worm, my ears attuned for the dusty, monotonous call of the yellow-hammer, trusted, respected and admired by all.

This whimsical reverie was shattered beyond repair by the cyclonic arrival of a customer, her nostrils flared, her eyes flashing, breathing fire and vengeance – and few creatures on this earth are more awe-inspiring than a housewife who believes that she has been done out of her rights. Faced with this terrifying manifestation of righteous wrath, I cringed, pulled my forelock and abased myself in the best tradition of the trade. After the first wave of anger had to some extent abated, I made so bold as to ask what the trouble might be. Still spluttering with indignation she managed to blurt out that, on opening a tin of rhubarb, she had found, of all things, a finger-stall, sodden with sticky pink juice, leering at her from within. This, as she rightly pointed out, was revolting enough, but, as she confided to me later, she couldn't stand rhubarb, anyway, and had only opened the ill-fated tin because she had run out of anything more attractive and her husband had turned up unexpectedly on leave.

What, she wanted to know, did I propose to do about the

matter? Controlling my temper as best I could, though seething with indignation at the injustice, I pointed out that, although blessed with many virtues, X-ray eyes were not included amongst them. I suggested she get in touch with the manufacturers. Only slightly mollified, she swept out of the shop and I did not see her for a week or so. When, in the fullness of time, she returned, I asked her what the result of her official complaint had been. This was an unwise move on my part. For a moment I thought she was going to smite me with her hand-bag. However, she managed to control herself and replied, in a voice so strained that it threatened to crack at every syllable — 'You suggested I wrote to the firm concerned enclosing the offending article. I took your advice. And what was the result? They sent me forty-eight tins of rhubarb! And if there is one thing on the face of this earth that I detest, it's flaming rhubarb!'

Another incident could have had more serious results. Some months before leaving London and smitten with a singularly tenacious attack of flu, which had reduced me to a pulpy apathy, I asked my doctor if he could do something about it. He gave me a prescription which I duly presented to the nearest chemist, receiving in exchange a small phial of pills. These pills proved to be of an attractive mauve hue and were shaped, for no apparent reason, like miniature hearts. Although I didn't realize it at the time, they were the famous 'purple hearts' which were later to figure so frequently in the press as scape-goats for the high jinks and misdemeanours of a whole generation of misguided youth. At the time, however, they looked innocuous enough and I swallowed one or two without thought of any repercussions.

Slowly and imperceptibly a feeling of supreme well-being crept over me. Gone were the lassitude, the shivering and the nausea, and in their place came a subtle sense of excitement, of anticipation of something indefinable that seemed to lie ahead. Above all, I was overcome by a feeling of brotherly affection for anyone, friend or stranger, whom I happened to encounter. It was like having drunk a considerable amount of alcohol but without the taste, the muzziness and the after-effects. Anyway, it certainly wiped out all traces of malaise, rapidly and permanently. As I still had half a dozen or so of these miracle-workers left, I put them in a suit-case and forgot all about them until they and I eventually found our way to Emsworth.

As the owner of a small country grocer's shop I frequently found myself playing the part of confidant, philosopher and friend to a wide circle of acquaintances. I was, I suppose, considered a bit of a *rara avis*, by no means everybody's idea of a conventional rural tradesman. I was comparatively young, apparently unattached, much travelled, and had arrived in their midst, not only with an eagle and various other assorted beasts, but with a certain aura of mystery, little of which I deserved.

Shortly before my first Christmas in Emsworth, a young woman whom I knew fairly well and who was a regular and comparatively liberal customer arrived one evening in a state of acute depression. She was suffering from a headache, sore throat and all the other horrors that indicate the imminent onset of disease. She asked me if there was anything I could recommend to allay the all-too-obvious symptoms. She was a nice girl and I liked her (and her custom), but not to the extent of wishing to share her bacteria, and so I suggested aspirins and all the other usual proprietary items that I could think of. She wasn't impressed and appeared to be disappointed in me as the dispenser of suitable ju-ju. Perhaps she expected me to leap about waving a skull on the end of a stick whilst muttering all manner of incomprehensible incantations.

Finally, wishing to bring the interview to a close and to ease her out into the night, I thought of those 'purple hearts' which had done so much for me. I gave her a couple of these harmless-looking objects and she went off happily enough, her faith in me restored. The following morning, shortly after opening-time, the door of the shop was flung wide and I was faced by an enormous and exceedingly choleric petty officer. 'What the hell,' he demanded, 'do you mean by half killing my wife?' I could only listen, blanching with terror at this torrent of nautical wrath. Eventually he calmed down enough for me to ask him exactly what he was talking about. It transpired that the girl to whom I had in my stupidity given the 'purple hearts' had gone straight home and swallowed them, with the result that she soon felt much better, so much better, in fact, that she decided at the last moment to go to a party. Here she accepted two gins and tonics, had drunk them happily, and had collapsed unconscious at her horrified hostess's feet. I explained my part in the unhappy incident. The PO, who was a reasonable enough fellow, accepted my apologies, bought a packet of cigarettes, and we parted on friendly terms – but I made a vow, there and then, that this would be the last time that I would play the part of medicine man to this community, or any other, for that matter.

Only a mile or so from the shop lay the western border of Sussex, the county that I had always considered *my* county, if only because it was the one that I knew best and where I had spent the longest and happiest period of my life. The great, rolling, grassy chain that divides the southern seaboard from the woods and farmlands of the Weald comes to its sudden, unexpected end at Harting Down, which is clothed in what I believe to be the biggest, oldest and densest natural yew forest in Britain. This is the haunt of herds of fallow deer, which are supposed to have escaped from parkland owned by large estates nearby but which I like to think are direct descendants of the deer which were roving the area free and unfettered when William Rufus fell mortally wounded whilst hunting in the New Forest, no great distance away to the west.

These fallow deer were not the only large wild beasts that trod those secret pathways. Pedalling homewards one evening, after a visit to some friends who lived close to the little straggling village of South Harting, I was thinking about nothing in particular other than the hope that my brakes would hold as

I skimmed down one of the long, steep, tree-shrouded hills that are so characteristic of the district. Glancing idly about me as I reached the foot of the hill I noticed a hole (probably a wood-pecker's disused nesting site) about twenty feet up the trunk of a beech tree by the side of the lane. It looked just the sort of site that a colony of noctule bats would take over as a day-time roosting place and I stopped to investigate, leaning my bike against the high mossy bank.

Failing to hear the shrill, strident squeaks that these rather quarrelsome beasts often emit even during the day-time when they are supposed to be asleep, I searched the ground beneath the hole for the bitten-off and discarded wings of Yellow Underwings and other large-bodied moths which often betray the presence of a noctule colony. There were no signs of those either. Vaguely disappointed, I seized the handle-bars of the bicycle and mounted to continue my homeward journey. I happened to glance to my right, where, beyond a scraggy hedge, a long, narrow field sloped steeply away to the woods, perhaps two hundred yards beyond. In a corner where fields and woodlands met stood a large chestnut-brown animal, which to my casual glance appeared at first to be a solitary cart-horse. The old-fashioned heavy farm horses are no longer a common sight and, being in no particular hurry, I climbed the bank to have a closer look.

What had seemed at first sight to be a Shire or Clydesdale appeared now, in the uncertain light, to take on the guise of a cow moose. This struck me as being an unlikely species to encounter in the heart of West Sussex, particularly as this was before the British countryside became peppered with Safari Parks and other commercialized 'wild life' establishments. I caught hold of a projecting ash stem to hoist myself up and over the low hedge, at the same time scuffling my feet in a crackling drift of last season's dried and withered holly leaves. Instantly the distant figure stiffened and swung its head to-wards me, one great trumpet-shaped ear cocked forward, the other backward, to ascertain whence the startling sound had come.

Much to my relief, I saw that it was not a moose but a magnificent red deer stag with but the rudiments of its new antlers showing like a coronet between the flickering, ever-moving ears. Unable to resist the temptation to see what it would do, I showed myself in full view. Instantly the stag turned and, with the delicacy and precision of a top-class show jumper, raised its forefeet and with a heave of its powerful hind-quarters cleared the boundary fence without a sound, disappearing like a wraith into the thickets beyond.

This was one of the few occasions on which I had no dogs with me but I couldn't help wondering what sort of show Tally would have put up if he had got the stag clearly in his sights. The saluki would have had neither the weight nor the strength of jaw to bring down such a powerful quarry, although he might have matched him in speed and stamina. Tally's pursuit of a wild, free and unwounded red stag would have been a contest worth watching. He would have enjoyed it enormously and it is unlikely that it would have done much harm to the stag either.

I do not consider the fact that I get a thrill from watching an animal fulfilling under natural conditions the purpose for which it was designed – which, in the case of a saluki or a bird of prey, is of course hunting – to imply that I am a bloodthirsty or unnatural monster. I do not intend to indulge in a long diatribe about the ethics of field sports. I do not shoot, I am no fisherman, but I do enjoy and will continue to practise the sporadic use of hawk and hound, separately or together, in pursuit of the quarry for which they have been developed since the beginning of time.

I have always thrilled to the sight of a saluki, graceful as a cheetah and almost as speedy, floating over the ground, feathered ears and plumed tail streaming in the breeze, desert eyes glowing with concentration and the excitement of the chase, whilst far ahead races the wily quarry, strong, fit and determined, knowing that the odds are weighed heavily in its favour, that sanctuary lies ahead, and that by its own courage, cunning and refusal to accept defeat it will usually win through to safety. On those few occasions when the dog does succeed in overtaking, outwitting and accounting for the quarry the end is quick and painless; far more merciful than the horrors of death by gunshot wounds, snare or poison.

The opponents of field sports believe, or say that they believe, that sportsmen revel in the death of their quarry. This

in my experience is far from the truth. In both falconry and
the sort of one-dog one-quarry type of coursing which I have
followed, the kill, if any, more often than not takes place out of
sight beyond the next hill. All the owner is likely to see when
he stumbles panting upon the scene is a triumphant hawk or
hound standing motionless above its quarry. There is no sign
of a struggle, no blood, and all around is quiet and still.

Think, for a moment, of another sport – one theoretically
free of the stigma of blood. A man stands on a step-ladder in a
narrow, house-flanked back-yard, eyes searching the far
horizon to greet an approaching dot in the sky. This tiny
speck, fighting its way homeward at fifty or sixty miles an
hour, swings closer, rapidly taking shape, flickering wings
half closed, and with a final slanting dive plunges through the
narrow opening waiting to receive it. After battling its way

down the length of the wind-swept Pennines or skimming the choppy waters of the Channel, one of his racing pigeons has arrived safely at base.

This pigeon, together with hundreds, maybe thousands, of its kind, was released hours before, perhaps from some sunlit station yard in Normandy. With a scrambling rush they hurled themselves into the air, quitting the wicker travelling-baskets with the impetus of greyhounds from the slips. Wheeling together, they moved into their long-distance flying formations and set off, a multi-coloured throng, blue check, gay-pied and mealy, impelled by what Ernest Thompson Seton has called the 'unswerving, God-implanted, mankind-fostered instinct' that leads them ever homewards, despite the ravages of hunger, thirst and weather. Over the sea, over mountain, valley and city, mile after throbbing, wing-wearying mile, never deviating, until at last the goal is sighted, a tiny white-painted loft, a mere pin-prick in the landscape. Without pause the wings are trimmed for the final descent and like a feathered rocket the pigeon, sanctuary gained at last, plummets through the narrow dowelled opening to seek the nest-box where its mate and young are waiting.

This is one case. Others there are that have a less happy ending. The number of pigeons that fail to make it to the home loft, that get blown off course, shot, lose arguments with telegraph wires or just disappear without trace, must be legion. For many years, I have played the part of host, temporarily or permanently, to innumerable racing pigeons. I keep a small assorted flock of so-called fancy pigeons (fan-tails, turbits, nuns and croppers) with a few rarer and more exotic varieties to add a touch of class. Each year during the racing season their numbers are augmented by stray homers that arrive unseen, appearing literally out of the blue. These birds are often in a parlous state, exhausted, emaciated, at the end of their strength. Some just manage to make it to my loft, arriving on the point of collapse. Some are too far gone after possibly days of wandering to do anything but sink into a coma from which only a few awake. The wastage amongst these pigeons must be colossal.

Yet I have never heard pigeon-racing described as a cruel sport. I have yet to read in the popular press of attempts to picket pigeon lofts or of hordes of angry bird-lovers attending the start of the long-distance races waving banners with the words 'Be fair to pigeons' writ large upon them. Personally, I doubt if a homing pigeon enjoys the actual process of finding its way home, particularly in the latter stages of the race or when weather conditions and a possibly faltering sense of direction combine to wear down the bird's sublime confidence. I would be surprised, in fact, if the birds are any happier than is the fugitive hare, freshly sprung from its form, strong, confident, with all its wits about it.

A pigeon race may last from first light to dusk; the stragglers, those that arrive, returning sometimes days after their release. A course, on the other hand, lasts but five minutes (ten is exceptional), the quarry being killed cleanly and quickly or, much more probably, making good its escape with no after-effects, except what it has learned from the experience. I am not condemning pigeon-racing, which, properly conducted, is a fine and rewarding sport. But it should not be forgotten that there is another side to it, apparently unconsidered by those who are loudest in their condemnation of any form of field sport.

High on the wooded slope to the south of the grassy hare-haunted valley I was resting, my back propped against the grey trunk of a huge oak tree, wrinkled and time-honoured as an elephant's hind leg. Below, out of sight amongst the bracken and elder scrub, the dogs were working. Tally was silent as ever and only the sudden and almost hysterical yapping of Bambi, questing the trail of rabbit or stoat, told me of the joyful activities of the two hunters as they revelled in the freedom of the immemorial forest land. Around them moved its teeming population of small, furtive, wild things, which were leading much the same lives as their ancestors had led when the oldest yew tree was but a seedling struggling upwards towards the light. From somewhere far off came the slow, satirical chuckle of a green woodpecker, followed shortly after by the challenging double-syllabled crow of a cock pheasant.

Not far below my resting-place, where the forest floor levelled out, was an almost circular open space, perhaps fifteen feet in diameter. It was just the sort of place that should have contained one of those shallow pools of rainwater that are so familiar a feature of those southern woodlands. There was, however, no rainwater — merely a fern-fringed mossy bowl, the home of a colony of bank voles, which were scampering about collecting fallen berries and other delicacies, their tiny russet bodies and round, blunt faces appearing momentarily as they followed each other up and down their own tangled, labyrinthine pathways.

A narrow trail that might have been worn by the feet of generations of badgers or foxes led up to the rim of this depression, fading away to reappear on the opposite side. Such tracks are common enough and I only vaguely wondered who or what was using this secret highway through the forest. From far down the valley came a sudden volley of barking, splintering the deep hush of the heavily wooded hillside. Came a pause, in which the silence seemed to close in almost tangibly from all sides, followed by a further burst of yelping, more urgent than before. Bambi's screaming war-cry was drawing closer.

Once more all was quiet. A minute, perhaps two minutes, passed. I thought I heard the squelching thud of something heavy approaching quickly. I stood up and stepped behind the trunk of the ancient oak tree, pressing close beside it. The cantering footsteps were coming up the trail on the opposite side of the clearing. They paused, but came on again, faster than before. The bushes parted with the dramatic suddenness of a rising theatre curtain, and on to the mossy stage bounded a magnificent fallow buck, in all the glory of his deep, rich brown, autumnal pelage, his wide-spreading palmate antlers held aloft like candelabra.

He was a superb animal, appearing in that natural setting twice the size of those seen lackadaisically browsing in sheltered ease behind the walls of some park. Standing there motionless and alone he looked almost primordial. Size apart, he put me in mind of the long extinct great Irish elk (itself only a huge primitive fallow). I had only a moment to admire him in all his untamed beauty. Turning to gaze in the direction whence he came he shook his head, gave a curious, sneezing bark, and with an elastic bound was gone, leaving only the swaying bushes to mark the way.

A few minutes later I heard the approach of galloping feet.

Out from the darkness beyond and into the clearing burst Tally. Tail high and nose to the ground, he was working out the line which his quarry had taken. It is generally believed that the greyhound breed of dogs, of which the saluki is an ancient and honoured member, have little sense of smell. Although as hounds of the open desert they prefer to use their marvellously acute eyesight, they can, when the occasion calls for it, follow a fresh scent with the best. I called Tally, who looked up, grinned, flicked his ears in recognition, and promptly turned his back on me, exactly as if he were saying, 'Don't bother me now, old boy – can't you see I've got a job to do?' After casting about like a professional foxhound he broke into his tireless canter and plunged through the autumn-tinted barrier of foliage, following exactly upon the trail of his long-vanished quarry.

Once more silence descended, broken only by the sibilant calls of a wandering party of goldcrests and coal-tits searching for chrysalids in the crannies of a black and rotten stump, victim of a lightning flash long years before. A pattering on the dried leaves, light as raindrops, and Bambi appeared, slender as a leveret but dauntless and determined as a dog three times his weight, the hunting fervour blazing in his prominent hazel eyes. How far he had run and what obstacles he had overcome, I knew not, but he looked as fresh as if he had just set out. Unlike Tally, he welcomed me with a flurry of affection before he too melted into the sun-dappled under-growth, blue-grey amongst the blue-grey shadows.

With the passing of the stars of the show, other lesser dramatis personae took their place upon the natural stage. A stoat, chestnut-coated and white-chested, undulated like a huge furry caterpillar through the tangled grass, sitting up, its fore-paws held meekly before it, its snake-hard eyes and twitching, whiskered muzzle questing for the scent of mouse or vole. Silent and elusive though he might be, he none the less failed to escape the obnoxious attentions of the local Mafia, in the shape of a pair of carrion crows, three magpies and a jay (who, with his warm fulvous plumage, unexpectedly tropical blue black and white wing flash and general air of having stepped out of some avian Savile Row, looked worthier of better things than robbing fellow predators).

The two crows, implacable killers of any small or helpless living thing that they encountered, dived at the stoat, snarling their raucous nasal challenge, stooping with surprising speed and dexterity for such apparently leisurely fliers. They dropped to within a few feet of where the small, fierce hunter crouched but feared to press home the attack. The magpies and the jay flirted their tails, flicked their wings, shouting abuse, whilst bouncing about in the lower branches like indignant rubber balls. The stoat, angry but unafraid, his ears and nerves sickened by the cacophony, switched his slender black-tipped tail and dived into an isolated shallow burrow that had once been the 'stop' wherein a doe rabbit had reared her family of short-eared, silky-coated youngsters. Deprived of their sport, the gang of noisy reprobates split up and went off to resume their prying, rapacious ways elsewhere.

First to return was Tally. He trailed in from the direction I least expected him and flopped down beside me with the air of one who has accomplished much, but is too modest to talk about it. His silky coat was plastered with sticky clay, his tail and ears a mass of spiky burrs and bits of dried bracken. He was panting and his sides heaved but after a long drink from a

convenient puddle he lay down with his head held high and his fore-paws stretched in front of him, making him look much like one of Landseer's Trafalgar Square lions. Watchful and alert, he seemed half inclined to set off on another foray.

The buck had had a long start and must have eluded him amongst the tangle of bramble, ferns and closely growing forest trees. Fallow deer are masters of the art of vanishing, of apparently just ceasing to exist. Unlike their compatriots, the wild red deer of Exmoor, they never, if it can be avoided, seek to out-distance their pursuers in the open by sheer speed and stamina. Rather they make use of every natural feature, sinking into the background, choosing the most impenetrable thickets as a barrier between themselves and danger, not emerging until all is safe and quiet once more.

It must have been ten more minutes before Bambi came prancing into view, looking as spruce and debonair as if he had just left his show bench at Cruft's. He had been hunting more or less non-stop for an hour and yet he showed no signs of fatigue whatever. It was clear that he had returned simply because he had lost the trail of his huge quarry, had lost touch with Tally, and was feeling lonely and vulnerable out there in the echoing vastness of the woods.

I shared a pile of ham sandwiches with the dogs before we made our way down through the chalky roadway tunnelled out of the downs and canopied by a screen of intertwining beech branches, which seemed to meet over my head like the swords of a guard of honour at a smart military wedding. I had hidden my bicycle beneath a pile of withered larch branches close to where the track lost its way amongst the leafy wooded rides and now I retrieved it, mounted, and we started the long slog homewards. The dogs had ten or more miles to travel but they settled into their rhythmic loping canter, stopping to sniff and to examine anything that might be afoot. When eventually we arrived at the village of Westbourne, a mile or so from Emsworth, they seemed almost as fresh and spruce as when they had first set out many hours before.

Chapter Fourteen

I shall always remember Westbourne for two reasons. Firstly, because it was skirted by a series of water meadows which contained the largest water-cress beds I have ever seen. This sea of crisp, light green leaves and tiny white flowers was presided over by a solitary water rail (at least I assumed that he was the only one because I never saw him with a mate or brood of downy black younsgters). He used to skulk at the edge of the cress beds, which were screened by a protective jungle of reeds, rushes and yellow irises into which he would scuttle when disturbed while probing for aquatic life with his long, slender, kiwi-like beak.

The other reason for which I am unlikely to forget West-bourne was that it included amongst its attractive, time-worn buildings, The Cricketers Arms. This was no ordinary pub, if only for the reason that the licencee, Morton Swinburne, was no ordinary landlord. I had been told about him by a mutual acquaintance, who not only recommended his professional hospitality but happened to mention that Morton was 'interested in animals and all that sort of thing'. The name had intrigued me because my mother's maiden name had also been Swinburne. The family being of Northumbrian origin, it was unusual to encounter it in the extreme south of England. I made up my mind to visit him at his hostelry at the first opportunity.

I was not disappointed. The public bar was at first sight much the same as thousands of others, all horse brasses and faded photos of what might have been Victorian darts teams or similar social gatherings, but, glued to the wall above the battered and much abused piano, was a poster yellow with age, which, if I remember rightly, bore words which ran something as follows: 'A fight will take place between Morton Swinburne of Westbourne and Charlie Cheney, the Chichester Chopper, over twenty rounds, to be fought on the village green for a purse of ten gold sovereigns. Winner to take all.'

Morton, as it turned out, had been one of the legendary bare-fist fighters and, judging from the number and variety of his trophies, an exceedingly successful one. He was also, and even more remarkably, a fellow of the Zoological Society of London. The incongruity of the two associations surprised him not at all. When Morton himself appeared from the back premises he seemed somehow to combine the two unlikely characteristics to perfection. Whereas, knowing of his pugilistic past, I was expecting a Goliath of a man, broken-nosed and cauliflower-eared, a sort of anglicized version of Victor McLaglan, the reality was quite otherwise. He was smallish, under five foot eight inches, but as hard and compact as a Staffordshire bull terrier and, despite being well over seventy at the time, in far better physical and mental condition than many men in their early forties. When he spoke it was with the gentle, precise tones of a member of the Athenæum.

After a short preliminary verbal skirmish, in which he was obviously weighing up my qualifications (as a naturalist, not, thank heavens, as a pugilist), he suddenly shot a question at me. 'Can you,' he demanded, 'give an example of parallel development without relationship?' This, apparently, had been one of the questions he had been asked when applying for fellowship of the Zoological Society, in the days when you had to know something of your chosen subject and it wasn't just a matter of forking out a substantial cheque to cover your subscription. I pondered for a moment before replying. 'The condor and the griffon vulture,' I suggested. 'No,' he shot back at me, 'you are wrong there. They are both birds of prey.'

Actually, though I didn't like to tell him so, I was quite right. The American vultures, which include both species of condor, are descended from some sort of primitive stork-like ancestors, whereas the true or Old World vultures are descended from prehistoric eagles, who presumably decided that it was easier to wait for something to drop dead than to risk life and limb and the expenditure of a lot of energy by going out and killing prey for themselves. However, I deemed it unwise to argue the point at this juncture.

'I'll give you an example,' said Morton triumphantly. 'The porcupine and the hedge-hog. Both have prickles and much the same general shape, but whereas one is an insect-eater, the other is strictly vegetarian. Come on – try again!'

After a long pause I came up with the swift and the swallow. Both have long narrow wings, spending most of their time in the air, catching insects in flight, but whilst the swift is distantly related to the night-jars, the swallows (and martins) are really specialized members of the passerine family. Morton accepted this, my honour was restored, and we downed a comradely pint together.

Morton's wife appeared and we were introduced. Never have I met a better adjusted couple. She had been a snake-charmer with a well-known circus at one time and still had a

153

passion for reptiles, which she satisfied to some extent by keeping a small vivarium containing green, wall and sand lizards, not to mention slow worms and tree frogs, on a ledge by the window in her private sanctum. She produced a press cutting from her days in show business, which portrayed her with what appeared to be the grandfather of all anacondas draped confidently round her slender shoulders. This monster had escaped from its travelling bag while she was on tour in some far northern town, which it had brought to a virtual standstill, until a frantic search had brought to light the rather anti-climactic reality that it had somehow managed to ease its entire length through a minute hole in the upholstery of the bed and was lying happily dormant amongst the springs, from which it showed the greatest reluctance to be extracted.

Morton had heard of Random (whom, for reasons known only to himself, he insisted on calling Broadbent) but he had not met her until one evening, when being visited by a fellow zoologist, he rang me up and asked me if I would bring her over for their joint inspection. This I did. It was a Friday night, both bars were packed, Random was in a fractious mood, and the ensuing chaos can better be imagined than described. That peaceful English pub took on much the atmosphere of a western saloon when the bad man strides in.

It so happened that Morton's step-son, a highly successful cartoonist, was amongst those present on that fateful night, and he, safely tucked away in an ingle nook, observed, made mental notes, and finally captured the scene for posterity. His picture, 'The Night of the Eagle', was a minor masterpiece. It duly appeared upon those smoke-blackened walls, dominating the aged prints of cock-fights, midnight steeplechases and other works of art, so much a part of such establishments. By guile, flattery and stealth I did my best to appropriate this unique work, but to no avail. For aught I know, it hangs there yet, silent proof of an event that Westbourne is unlikely to forget for many years to come.

The high spot of the night came after the last customer had

departed singing into the darkness. Morton and I, together with a favoured few of his special friends, stayed on for a final pint or two. Random, now hooded and serene, perched motionless on the back of a chair, oblivious of the conviviality going on around her. From bar counter almost to ceiling arose twin columns of copper coins, destined eventually for some charity. They were, I suppose, some three feet apart and as solid-seeming as a pair of Doric columns.

The beer flowed and comradeship prevailed. Someone — I forget who — was unwise enough to bet me that Random would not be able to fly between these columns without knocking them down. Full of beer and optimism I accepted the challenge and the wager was on. The chair, with Random still sitting upon it, was carried behind the counter. I unhooded her and removed her leash. Then I returned to the saloon and, holding up my gloved left fist, called her urgently by name. She craned her neck, peering about her in the smoky haze. She picked me out, tensed her muscles, half opened her wings, closed them again uncertainly, bobbed her head to judge the distance, and finally left the back of the chair with one great elastic bound, coming straight for the security of my fist, the perch she knew so well.

For a moment I thought she was going to make it. She is a genius at manœuvring her way through seemingly impossible openings. Maybe she was confused by the artificial light, the lateness of the hour, or the general air of expectancy; anyway she misjudged her distance. One wing butt shot out, catching the left hand column about half way up with all the weight and speed of her body behind it. Samson's last fling at the Philistine capital could hardly have matched the destruction and the desolation that followed. The sturdy pile of coins fell like an axed sapling, scattering its accumulated wealth broadcast. It must have taken a good half hour of grovelling, groping and swearing before the last fugitive penny was retrieved. All that was left for me to do was to hang my head, pay up my ten bob, order a final round in an attempt at appeasement, and hurry out with the confused and indignant Random into the soothing anonymity of the night. It says much for the good humour and tolerance of the Swinburnes that they remained amongst my closest acquaintances until I finally left the district for ever.

When it became generally known that there was an eagle in the district my premises became a sort of Mecca for hordes of would-be falconers, practising falconers and ex-falconers. I had no idea that such an interest in raptors existed. Hardly a week passed without someone arriving on my doorstep proudly carrying a kestrel, goshawk, lugger falcon, or something even more exotic and improbable, perched statuesquely on his fist. About this time the Country Landowners Association had started running their Game Fairs. These are sporting get-

togethers, held annually on large estates all over the country. In addition to the usual rural events, such as gun-dog trials, fly-fishing competitions and clay-pigeon shooting, they began to introduce displays of falconry, in the course of which trained hawks were flown to the lure. There was also a number of birds of prey on exhibition, sitting on block or bow perches on their specially prepared weathering grounds. These displays were provided by the lordly and long-established British Falconers Club – the legendary BFC – which had at one time or another included in its ranks such doyens of the sport as Colonel Faithful, Guy Aylmer, the artist George Lodge, and the author T. H. White, who was responsible for such immortal works as *The Goshawk* and *The Sword in the Stone*.

The result of all this publicity was that numbers of people who had never heard of falconry or, if they had heard of it, assumed that it had gone out with the invention of the crossbow, now became determined to try the thing for themselves. The animal dealers were quick to latch on to a hitherto untapped market. Whereas previously they had confined themselves to selling black bear cubs to mews dwellers in Belgravia, their advertisement columns in the livestock papers now held out entirely new prospects. 'Egyptian Vultures and Pariah Kites especially selected for Falconry.' 'South American Crane Hawks. Would make Ideal Falconer's Birds. £50 each.' They were on to a good thing and they intended to make the most of it whilst the craze lasted.

I myself fell a victim to this sort of thing. An advertisement appeared in a highly respectable agriculturists' paper offering an Imperial eagle for sale at a reasonable price. I had always had a nagging wish to own an Imperial eagle, if only for the reason that I knew little about these birds, apart from the fact they were rare, beautiful and looked vaguely like golden eagles. I sent off my cheque and in due course the station rang up to say that a box of livestock had arrived. The box looked much too small to contain anything as massive as an Imperial eagle but none the less I felt the glow of excitement which I always experience when anything new and alive arrives at my door. Fumble-fisted, with poorly-controlled zeal I tore off the hessian covering and peered in.

Glaring up at me from challenging ice-blue eyes, snowy-breasted and mottle-backed, was, as I suspected, no Imperial eagle, but an immaculately beautiful, arrowy, arrogant and altogether entrancing young changeable hawk eagle. I disengaged her talons from the filthy canvas floor, pulled her out, getting savagely clawed in the process, and perched her upon the back of a kitchen chair in the darkened room. I named her Razmak on the spot, in the belief that this could possibly have been her place of origin. Although secretly delighted, I saw no reason to let the dealer get away with this sort of deception, and so I wrote a mildly abusive letter, stating that if I had wanted a hawk eagle I would have ordered one and suggesting that if he didn't know one eagle from another he should buy a book on the subject and swot it up! This was long before the

passing of the Trades Description Act but his reply, when it came, was so abject that I felt almost sorry that I had written.

Whatever the cause of this sudden upsurge of interest in anything connected with birds of prey, it was directly responsible for my meeting a couple who were to become among my oldest friends. It was an excessively hot Sunday afternoon and I, having nothing of great import to deal with, was lying stretched upon my bed perusing the pages of Whistler's *Birds of India* to find out what he had to say about the eagles of that country in general and the changeable hawk eagle in particular. I heard a knock at the back door and ignored it, believing it to be caused by someone wanting a packet of cigarettes or a loaf of bread. The knock came again louder than before.

Scowling and muttering, I clattered down the stairs and opened the door, preparing to offer a few pithy and offensive remarks about the sanctity of the British Sunday afternoon. A young couple with a small boy were standing smiling up at me from the foot of the steps that led to the back door. I grunted some inquiry as to what the heck they wanted and was ashamed of my rudeness when the man replied that he was a practising falconer, had an eagle of his own, and would like to have a look at mine. I ushered them in and introduced them to Random, who, I regret to say, was even less welcoming than I had been. Thus it was that I met the Mussared family: George, Hazel and Stephen. Years later, George confided to me that never had he met such a brusque, surly and offensive individual – a description which I regret I was quite unable to refute.

After this unfortunate introduction we quickly discovered that we had much in common and became close and enduring friends. George was then serving in the navy and had a temporary shore job, being stationed at Haslemere, in the centre of the pink gin and tennis party belt, a situation which pleased him not at all, as he detested pink gin and wasn't overenthusiastic about tennis either. At the time he had a tawny

eagle named Thor and he lost no time in taking me over to his house to meet him in person.

George was and still is a fanatical falconer, and one of the very few people I have met who has had any success with tawny eagles. Almost without exception the ones that I have had the misfortune to encounter (and that includes one of my own – an African tawny named Juba) did little except sit lumpishly on their perches, chirping in a thoroughly un-aquiline way and biting pieces out of one's face if they got the slightest chance. Thor, on the other hand, flew well, and would come to George's fist or lure over long distances. He was, in fact, though I hate to admit it, at times a good deal more obedient and co-operative than Random herself. Over the next few months the Mussareds and I came to know each other well. George's shore job was coming to an end and he was about to be posted to, of all things, HMS *Eagle*. However, before that event took place, we used to meet as frequently as duty, distance and the price of petrol permitted.

An ever-growing band of enthusiasts would meet to discuss endlessly the topic that had brought us all together. It was generally agreed that there was room in Britain for another falconry club, if only so that keen but uninitiated would-be falconers might learn from those with more experience the rudiments and ethics of the sport: that, for instance, a rabbit hutch is not the ideal off-duty home for a red-headed merlin, or that he who loses a hawk with leash and swivel attached is as worthy of a black mark as those who shoot a fox in Leicestershire or gaff a spawning salmon in the Spey. Thus, after much debate and a good deal of paper work, the Southern Hawking Club was launched. This club soon became so well known and received so many applications for membership from all over the country that before long the name was changed to the Hawking Club of Great Britain. As such it was known for several years; until, indeed, its amalgamation with the BFC.

Chapter Fifteen

I was listlessly turning over the pages of my morning paper on the first Monday morning of March, 1965. It was as dull and depressing as usual – as dull and depressing, in fact, as the weather outside. I propped the paper against the coffee-pot and was toying with the idea of tearing a strip off to use as a spill to light my cigarette when I noticed a photograph of what looked vaguely like an eagle perched amongst the wintry, leafless branches of a tree. Interest awakened, I looked closer. It was indeed an eagle and beneath the picture was written, 'The Golden Eagle which escaped yesterday from the Zoo resting in a tree in Regent's Park.'

Thus began the escapade that was to last for nearly a fortnight, to hold enthralled the imagination of Britain and most of the civilized world and, incidentally, to produce some outstanding examples of journalistic extravagance, of which 'the spirit of the Roman legions lives again in Regent's Park' is only one comparatively mild example. How exactly the eagle (promptly dubbed 'Goldie' with a singular lack of imagination by the press and other media) escaped in the first place is not clear but he made remarkably little use of his freedom while he had it.

The bird was a male which came, I believe, from Finland as a young untamed adult. No subsequent attempt was made to tame him and, although he shared a cage with a large and equally unfriendly female, they showed little interest in each other and so could hardly be considered as a mated pair. Given this, there appeared to me to be no particular reason why he should remain in the district. I confess that I had every hope that his natural instincts would assert themselves and that he might float away on a current of air. However, although shadowed by embarrassed keepers and pestered by the public at large, he showed no such inclination and remained within half a mile or so of the Zoo boundaries.

Various abortive attempts were made to re-capture him, including a nocturnal raid by the London Fire Brigade who planned to dazzle him with lights at his roosting-place and grab him before he could pull his wits together. The only result of this was that, seeing and hearing the firemen's ladder approaching, he took off into the darkness, being reported next morning sitting on the roof of the pavilion at Lord's Cricket Ground. After several days fasting (no great hardship for an eagle) he at last bestirred himself, slew a muscovy duck which he stooped upon near the lake at Regent's Park and devoured half of it before being disturbed and driven off. Unlike Random, he lacked the courage and determination to fight for what he had caught and thus, half a crop to the good, he took to the trees once more. Each day the newspapers, radio, and television cameras reported his activities and showed some remarkable shots of him in flight and of the even more remarkable crowds who came to see him putting on his rather limited display of aeronautics. It was all good clean fun while it lasted. I followed his career with interest and contemplated, and dismissed, various plans to bring the circus to an end as speedily as possible.

I was relaxing in my sitting room before a blazing fire while the dogs lay around it in attitudes of abandonment when the door reverberated to a volley of urgent blows. I sprang up thankfully, grateful for anything that would break my present train of thought. I was not in the least surprised to see three fellow falconers standing heavily overcoated on the threshold. Had they been wearing cloaks and wide-brimmed black hats, and carrying sticks of high explosive and a pass key to the Houses of Parliament, they could hardly have looked more conspiratorial. Bill was brandishing a dead hare in one hand and a bottle of Vat 69 in the other. Peter had a huge net,

163

fastened to a bamboo frame, slung over one shoulder, whilst John was locking the door of a serviceable-looking safari-wagon. Altogether, it was a sight to rejoice my heart. 'Well,' we asked in unison, as we unscrewed the cap from the bottle of whisky, 'when do we start?'

The plan was simplicity itself. After a few hours' sleep, encouraged by the Vat 69 (the level of which was rapidly receding), we would rise and, with Random, the hare and the net in the back of the truck, rapidly descend upon Regent's Park as the first pale hint of a winter dawn was lighting the skyline to the east. Using Random and the hare as decoys, the net, cunningly manipulated by strings from a distance, would be held ready to embrace the hungry fugitive when he came down all unsuspecting to enjoy the first decent meal he had seen for days. What we intended to do with Goldie when we had extracted him from the net was the subject of much discussion. The legal position was not entirely clear.

The golden eagle is a wild bird, indigenous to Great Britain (or at least to a small and comparatively remote part of it). The fact that the bird had escaped and had made no attempt to return and that it bore no means of identification or proof of having been confined were points in favour of the argument that in law it was no longer the property of any individual or organization, and came, presumably, under the auspices of the Bird Protection Acts. I vaguely remembered a case in which someone lost a trained peregrine falcon which was eventually recovered by someone else, who refused to surrender it to its original owner. The case came to court and I believe it was decided that such a bird being *naturae ferae* (or something equally abstruse) it could only legally be regarded as someone's property when under the control of that person. The possessor forfeits his right if the bird reverts to the wild or is lost in the field, thus differing from a domestic animal like a ferret, which apparently belongs to its original owner wherever it is and whatever it may be up to (the owner being responsible for any damage the brute may do). All this seems logical enough – provided, of course, that it isn't one of my birds that is involved.

We had thought up some rash and romantic ideas of either holding the bird for ransom for vast sums of money or making a dash for the highlands of Scotland and releasing it there in the hope that it might become established and perhaps eventually mate with one of its Scottish contemporaries. My three confederates settled down for an uncomfortable few hours on settee, chairs and the floor, dragging themselves awake at half past four to gulp down cups of tea and slices of revolting-looking buttered toast. I left a note for my assistant, leaving the shop, the dogs, and everything else in his care, and, with Random perched happily in the back and the rest of the apparatus stowed beneath the seats, we slipped out and away on the seventy mile dash to London.

Thanks to the constant bulletins we knew more or less where to find our quarry but we had not bargained for the British winter dawn at its most unco-operative. The best-laid plans of even the most altruistic of eagle-nappers are as likely to gang agley as those of mice or men. The expedition was doomed from the first. We ran into black ice, white ice, and just about every shade of ice in between. We were blinded by sleet and bombarded by hail and finally, in some outlandish outpost on the western boundaries of Greater London, we lost our way, ending up in what appeared to be a knacker's yard. St Hubert, the patron saint of hunters and therefore of falconers, was for once most certainly not on our side.

Despite all this, we did not for a moment consider calling off the operation. We were fully committed and, come hell and high water, we were determined to carry on, if only because I had decided that a second golden eagle of the opposite sex might be a desirable acquisition. No one, to the best of my knowledge, had succeeded in breeding these birds, but there is always a first time and if anybody was going to do it, it might as well be me. I hadn't consulted Random on the subject but she was by now just about the right age and was certainly in perfect physical condition. This seemed as reasonable a way to acquire a male for her as any other. What was far more to the point, it was an adventure. We all felt we could do with something like this to break the monotony of life in the nineteen sixties. We felt pleasantly piratical. The idea of appropriating this eagle that had been hogging the headlines from under the noses of the watchful and, according to the radio and television reports, exceedingly vociferous zoo staff, not to mention thousands of self-appointed eagle-watchers, appealed to us immensely.

When we eventually arrived at Regent's Park we were so late that the whole expedition took on an air of anti-climax. In the first place, we were nearly arrested on arrival – a contingency for which none of us had bargained. I suppose, looking back, that we must have looked a pretty bunch of desperadoes, festooned with nets, slung about with defunct members of the leporidae, and dressed in garments more suitable for the Dogger Bank than one of London's Royal Parks. However, when we explained our mission, the park keeper became much more amenable, if not actually an active participant in the proceedings. We had arrived an hour or more behind our original zero hour, to find that ever-increasing hordes of sightseers were already pouring into the park from every direction. Our confederate the park keeper pointed out a posse of his zoological opposite numbers, who were clustered round the foot of a large tree.

About forty feet above the ground, motionless and splendidly aloof to all the hubbub below, sat Goldie, peering down at the throng with magnificent contempt. Indifferent to all the fuss, he turned his head and began casually to preen his scapula feathers. Even I, blasé from seeing my own eagle behaving daily in just such a manner, couldn't help feeling a surge of admiration and sympathy. Why hadn't the fool cleared off and sought the solitude of open country that lay within easy eagle flight to the north and west?

It was obviously beholden on us to do something to justify

our presence and so I returned to the parked van and brought out the hooded Random, who, I was secretly delighted to note, not only dwarfed the distant Goldie but appeared to have far better plumage too. Even from where I stood I could see that the refugee's tail feathers seemed worn and frayed. Since the whole situation had changed, we explained that we had driven up from Hampshire at considerable inconvenience and colossal expense to put ourselves and our experience at the disposal of the zoo authorities. These gentlemen now gathered in a huddle, obviously discussing us and, from the general look of them, none too pleased at our intrusion. Meantime, the attention of the onlookers had turned to Random, who now, unhooded, was sitting unconcernedly on the block we had brought with her. Cameras clicked, reporters reported, and small boys asked for feathers.

Before long we saw a deputation of zoological top brass approaching from the direction of the menagerie. They were led by one who, although we were not formally introduced, I recognized as Sir Solly Zuckerman, who at that time virtually owned the zoo, or at least took a big part in the running of it. He was exceedingly affable and after thanking us for coming, more or less gave us *carte blanche* to do as we thought best. It was obviously impossible for Goldie to remain permanently confined to the rolling but restricted acres of Regent's Park. There was unlikely to be enough food for him and, judging by the present assembly, he would have had little peace or opportunity to go out hunting for it. Furthermore, it was only a matter of time before he got into trouble. He had already shown a propensity to attack small dogs and apart from this, there was every possibility that he would eventually crash into an electric cable or come to a sudden end in some similar manner. By and large he would be better off in his aviary, where at least he would have regular meals.

We set up the net, pegged out the hare, and tethered Random by a long leash to her block, while some enormous tough who had appeared from nowhere took upon himself the job of keeping back the crowds. No one argued the point with him. Random looked up at Goldie, defying him to come down and rob her of one of the hare's hind legs at which she was contentedly pulling, utterly oblivious of the presence of what appeared to be the vast majority of the Greater London population. Goldie looked down, thought better of it, and swung out of his tree, disappearing in the direction of Primrose Hill. The show, as far as we were concerned, was over.

With the disappearance of Goldie the crowd stampeded down to have a closer look at Random, who, being Random, tore a sizeable piece of trouser-leg off someone who was imprudent enough to try to take liberties with her. As a final gesture and to show that Random was as much of an eagle and just as capable of a flying display as Goldie, I called her several hundred yards to the lure, right over the heads of the somewhat demoralized crowd and only a few feet above them. They had had their money's worth and so had we. Followed and impeded by hundreds we returned to the truck, settled Random on her block in the back, and roared off westwards. It might have been unsuccessful as an eagle-hunting expedition but it had certainly been an experience for all those present, as the national newspapers reported next day.

When we reached Cobham we decided to have a well-earned pint and to mull over the day's events. We left Random clearly visible behind her glass panels and strode into the nearest pub. We were half way through our beer when we saw the landlord talking to a couple of obviously regular customers. He called out, 'Is that an eagle in the back of your truck?' I replied, 'Certainly it's an eagle – a golden eagle. Actually, we've just come back from Regent's Park. The Zoo have lost an eagle, didn't you know?' He gave me an appraising look and disappeared behind the bar. We finished our drinks and put the empty glasses back on the counter. We were just in time to hear a voice obviously talking to someone at a considerable distance. The words were muffled behind a glass door but not too muffled for us to hear them. 'Is that the Zoo? Well, there are four blokes here, and I think they've been and gone and pinched your eagle!'

Although we returned from London with the same number of eagles as we had set out with, it had been an experience that we would not have missed. The following morning the daily papers were full of pictures of Random, posing in all her glory and taken from every conceivable angle. She even appeared briefly in both the BBC and the Gaumont-British news reels and this resulted in our being asked to appear in several television programmes, where, unlike on her previous performance, she behaved impeccably.

Not long after this escapade, and after Goldie was safely ensconced in his cage once more, I received a telephone call from London ITV. Someone, I never discovered who, had put my name forward as a possible competitor in the 'Double Your Money' programme. Apparently this anonymous sponsor of mine, tired of listening to an endless succession of experts on football, fishing and similar rather overdone subjects, had decided that it might make a change to have a falconer on the programme and, without the formality of consulting me, had sent in my name. The programme producers, apparently thinking much the same, now rang me up to find out if I was prepared to take part.

I hesitated, because I wasn't sure whether it would be good for the so-called 'image' of the sport to take part in this sort of programme; also I wondered what those haughty pundits, the long-established practising falconers, the members of the BFC, would think of it. However, I wasn't a member of the BFC and I didn't consider that answering a few questions about hoods and falconers' knots could do any lasting harm. Furthermore I could do with the money if I was at all successful. After all, £32 was not to be sneezed at and the possibility of winning £1000 was an almost unimaginable dream of El Dorado. I agreed at least to attend a preliminary interview which was to take place a week or so later at the Pendragon Hotel in Portsmouth.

This was an enlightening experience. There were a dozen or more hopeful potential participants (none of whom, I noticed, appeared at the real show in London). I was asked a few simple questions to see if I knew anything at all about the subject and was then told that I was a certainty for the real thing. The fellow whose job it was to weed out unsuitable contestants asked me, apparently as an afterthought, if I would be prepared to bring a real live bird of prey to the London studios. Without thinking, I agreed. About a week later the producer rang me up to tell me that the recording would take place the following Thursday and that under no circumstances must I turn up hawkless.

Well, I thought, the viewers might as well have their money's worth, but which of the birds should I take? I would be travelling up by train and across London by Underground. Random's box was a bit too cumbersome and unwieldy for this sort of thing. Besides, Random herself was becoming smug and self-satisfied after all her recent publicity. She was beginning to suffer from over-exposure. It was time she took a back seat for a bit. Bianca was too exotic-looking. She was not in fact a typical bird of prey, being exaggeratedly coloured and with such large feet and beak, both somewhat out of proportion, that she gave the impression of being something of a caricature. The obvious choice, therefore, was Razmak, the hawk eagle.

Razmak was the right size (twenty-seven inches long) and weight (three and a half pounds). Her pure white breast, blue eyes, and beautifully mottled upper parts combined to give an impression of quiet efficiency. With her streamlined body, her long tail and powerful legs, her huge feet and small dainty beak, she was indeed the raptor *par excellence* – photogenic without looking too dramatically glamorous. Also she was tame to the point of being virtually unflappable. She would fly long distances to my fist with a minimum of delay. She was quite at home under artificial lighting (her early training having taken place indoors during the long quiet winter evenings after the shop was closed for the night). She accepted strangers with the grace and dignity of a crown princess and was used to travelling unhooded in all sorts of transport.

As the show was a recorded one (not being transmitted until weeks after the departure of all the dramatis personae) I was asked not to mention the fact that I was taking part in it. I travelled up from Emsworth station with Razmak sitting imperturbable as a stuffed bird on my left fist. I carried an extremely ornate hood in my pocket, but only for use in case of emergencies, which never arose. In order to avoid undue publicity and thus to try Razmak's nerves too high, I travelled first class (I might add that my expenses were fully paid). None the less, at each station, and the train stopped at every one without exception, passengers came and went. I buried my face behind the pages of the only falconry book that I possessed in order to mug up on the subject as best I might during the last few hours left to me before the inquisition.

The British are a curious race. No one seemed the least surprised or perturbed at the prospect of being closeted for a couple of hours with a large, imperious and even lethal-looking eagle. The ticket collector chatted to us, admired Razmak, and to my intense relief made no suggestion that she should be provided with a ticket; nor did he, or for that matter the other passengers, ask me why I happened to be travelling to London with such an unusual companion. At Waterloo I hurried through the barrier, again escaping from being challenged for an eagle-ticket.

At the studio I was welcomed by the producer who took us to our dressing room and introduced us to Monica Rose and Anne Williams, the hostesses on the show. We were taken on a tour of the studios and shown all the technicalities, of which I understood just about as little as Razmak did. We posed for photographs with Monica and Anne and all had tea together in the canteen at the back of the set. Here I met John Ridgeway and Chay Blyth, the two paratroopers who had recently completed their transatlantic rowing expedition and who were appearing as celebrity contestants. John Ridgeway confessed to me that nothing he had experienced on that epic voyage had frightened him half as much as the ordeal about to begin in front of the cameras. This was a sentiment which I heartily endorsed. I had not rowed the Atlantic but I had fought a one-round contest with a buffalo and at that moment would gladly have done so again if I could thereby have avoided the appalling display of stage fright and general incompetence which I felt sure I was about to exhibit.

A stiff whisky at the studio bar did much to cheer me up, and when the time came for me to take the floor it was with the sort of numbed fatalism that I assume prisoners on the way to the block must have experienced that I shuffled through the curtains and into the blinding glare of the arc lamps. These seemed to me more like a battery of search-lights and even Razmak, accustomed to all the power of the eastern sun at its height, regarded them as a bit too much of a good thing. My hopes of untold wealth had already been shattered, because the contestant immediately before me (another angling expert) had got through to the 'treasure trail' and the most I could hope for was the £32 which would be mine if I answered the six questions correctly.

The first of these questions was, as always, a joke designed to put the contestant at ease, in my case so excruciatingly fatuous as to merit oblivion. Hughie Green, the quiz-master, was, I think, a bit put off his stride by Razmak, who failed to respond to his rather nervous advances and appeared as little amused as Queen Victoria is alleged to have been when confronted by an earlier comedian. I answered my questions, which were hardly as erudite as those one might expect from Bamber Gascoigne in 'University Challenge', and I duly pocketed my £32 and departed into the clamour and bustle of the London night.

Once the show was over and I could relax I realized that I had enjoyed the experience. Everything had gone well. Razmak had behaved in exemplary fashion and I had at least not returned empty-handed, which after all was what the exercise was all about. When I saw the show transmitted a few weeks later, sitting in my own room surrounded by friends and with a well-charged glass in my hand, I had the remarkable feeling of actually being in two places at the same time. Furthermore, I wondered if my voice, which I had never in point of fact heard before detached from my body, could really be as clipped and abrupt as it appeared to be, rasping at me from the depths of that ridiculous seventeen-inch screen. Of one thing I had made up my mind — a career in television, even if the opportunity presented itself, was definitely not for me.

Not long after this I was walking with the dogs towards the grassy, marshy peninsula that juts out into the sea to the east of Langstone Harbour. I paused before crossing the main Portsmouth road and at that moment Tally spotted a hare lolloping at the edge of a field of turnips at the opposite side of the road. Despite being nearly fifteen years old his eyesight was good, his hunting instinct unimpaired. He lunged forward with all the weight and power of his sixty-pound body behind him and jerked the lead from my hand as he plunged across the road. A car caught him broadside on and he was hurled into the curb. When I reached him he was dead. Fifteen years of friendship were over and my last link with the Africa I had known and loved was snapped. I borrowed a spade from a nearby cottage and buried him in the shade of the great hawthorn thicket where he and Bambi had so often hunted real or imaginary quarry together. As I walked home sick with misery I thought of Kipling's poignant verse. Why, oh why, had I once again given my heart to a dog to tear?

And yet — was it so tragic? Tally had died instantly, before he grew too old to enjoy the life for which he was so suited. He could run and hunt, his eyes and nose were both in full working order, and he was as fit and nearly as agile as he had been when he was six or seven years old. Surely that was better than to linger on with failing health and diminishing powers until

all interest had gone and only the vet's lethal injection remained as a final gift of a merciful death too long deferred. I
am not unduly sentimental. I have an open mind about the
possibility of a world beyond the one in which we live – and
yet for weeks after Tally's death I felt his unseen presence
close to me, and when I went for walks with the other dogs I
sensed, as indeed they seemed to, that we were not alone, that
Tally, silent and invisible, was trotting along with us. This
may be imagination or wishful thinking but the effect was none
the less exceedingly comforting and, talking to others who
have lost well-loved dogs of their own, I have been told that
they too have had the same sort of experience. There may
well be those who will dismiss this thinking as complete
twaddle or the worst form of anthropomorphism but they will
not, I hope, be reading this book.

Chapter Sixteen

Random slept in the empty garage at the end of the garden. When she wasn't sitting on her block or on the back of a chair in my sitting room she would retire, there, especially if the weather was so appalling that even she, who loves the wind and rain, could stand it no longer. She had a perch, a long thick branch that stretched across the otherwise empty building from corner to corner. There was also an upturned packing-case close to the second door, the one that opened on to the street. She used this packing-case frequently as a change from her more conventional perch. She found, no doubt, that the flat surface was the next best thing to a rocky boulder and as restful to her feet. Sometimes she would lie stretched out on it, her wings hanging down on either side, her head resting on the wooden ledge, in an attitude of utterly un-aquiline repose.

The daylight hours were normally spent in the garden, where she devoted much time and energy to the destruction of cardboard boxes, which were among her favourite toys and on which she would unleash her pent-up energy on days when she was not taken out for flight. It took her remarkably little time to discover just how to untie a falconer's knot and, having done so, she would seize the leash by the button at the end and pull it through the slits in the jesses. She would do this hand over hand, or rather hand over beak, with commendable single-mindedness until the whole of the eight-foot length had been released, whereupon she would cast it aside. After a few moments spent apparently in working out an itinerary, she would unhurriedly take wing, probably landing first of all on the garden fence, where she would bob her head and half open her wings before finally, her mind made up, rising almost vertically into the air and disappearing over the roof-tops of the row of terraced houses on the opposite side of the road. These aerial surveys of her domain were generally of a circular or semi-circular nature and, after a few minutes, she would come swinging into view, dark and purposeful as a Wellington bomber, from the opposite direction to that from which she had set out.

She seldom got into trouble during these periods of truancy, although on one occasion I had a frantic telephone call from an indignant householder, who, almost inarticulate with emotion, begged me to retrieve my eagle, which, he swore, was not only sitting on the kitchen table consuming the family's Sunday joint, which she had waylaid on its way to the oven, but was refusing to allow anyone into the room to retrieve the succulent and expensive haunch of beef. When I arrived, hood in hand, a few minutes later I found the family clustered in the passage whilst sounds of rending and tearing worthy of a pride of lions came from behind the half-open door. When, scarlet with embarrassment, I burst in upon her, Random, her mouth full of prime beef, had the effrontery to challenge my authority, drumming the table with mantling wings, throwing back her head and daring me to come between her and her unrightful booty. Normally I would have hesitated before intervening when she was in such a mood, but my embarrassment at her behaviour overcame my respect for her talons, and I thrust my gloved fist towards her. She left her ill-gotten meal and sprang upon my fist, squeezing with all her power, almost paralysing my forearm. Luckily I was able to grab her jesses. I rammed her hood on to her head and after a hasty apology, very well received considering the price of meat even in those far-off days, I hurried out and away. This was the only time that Random really disgraced me.

During the years 1965 to 1966 the new upsurge of enthusiasm for falconry reached its peak. Falconers and potential falconers of widely differing degrees of competence seemed to appear in practically every town and village in Britain. The consequent demand for anything that could conceivably be considered a bird of prey was almost insatiable. Dealers prospered and hawks suffered at the hands of some who were hardly capable of the correct management of an aviary full of budgerigars. Falcons appeared in television advertisements in connection with anything from tailoring to tobacco and thus the desire to possess these lovely birds grew ever wider.

As is so often the case where money is involved, the criminals moved in. A rash of hawk thefts broke out. Two peregrines were stolen from one establishment, several lugger falcons and a saker from another. Finally, a pair of golden eagles disappeared from a zoo in the Midlands. This made me think. Anyone who could kidnap a golden eagle must either be exceedingly courageous, exceedingly stupid, or an expert at the art of handling birds of prey egged on by avarice – an uncommonly dangerous combination. I was often asked how much Random was worth and I would reply: 'Well, actually, she wouldn't be worth a penny to you in the normal way, because you wouldn't be able to handle her, although, should you find anyone unwise enough to buy her from you, the price might just about cover the expenses incurred during your sojourn in hospital.' I wasn't bluffing. I had seen what she could do.

I had once asked a friend of mine, a keen and experienced falconer, to help me renew her jesses. He in his innocence agreed. Random was hooded and, as always under these conditions, perched rock-like and apparently relaxed on the back of a chair. I removed the old jesses, during which operation she appeared indifferent to what was happening around her. My friend advanced with one of the new ones and began to put it in place around her left leg just above the foot. Sensing the strange presence, Random tensed almost imperceptibly. My friend began to slip the point of the jess through the appropriate slit. As he did so she without warning struck forward with the speed of a fighting stallion. My friend was quick, his reflexes in perfect working order, but none the less, before he could withdraw his hand the few necessary inches Random's deadly hind talon caught his thumb close to the nail. It was a glancing blow, hardly penetrating the flesh, but within minutes his whole hand began to swell until it looked like a young melon. He had to go to the outpatients department of the local hospital for a series of penicillin injections and the pain did not entirely disappear for some days. I need hardly add that from that day onwards I changed the jesses myself.

Random has another curious idiosyncrasy. She is a very light sleeper, so light that in all the years that I have been owned by her I have never actually caught her in the conventional attitude of repose. Furthermore, she can apparently see nearly as well in the dark as a cat and if anyone, even myself, enters her mews at night without at first calling her name she is likely to attack as soon as the door is open. In addition to all this she shows an unexpected propensity to attack straight down the light path of an electric torch, in the same way that certain German pilots of the last war were known to dive their aircraft straight down the path of the British searchlight that was holding them in its mercilessly blinding beam. Fortunately I learnt early of this trick of Random's and made a point of never entering her quarters after she was supposed to be roosting without making sure that she knew just who it was that was disturbing her slumbers. As a final precaution I never put her away at night without first removing her leash, just in case it should catch on some nail or other unseen projection. There were, as I have said, two doors to the garage where she roosted. One opened on to the garden and the other the street. The latter was secured by a heavy padlock and was never normally used except when I wanted to scrub the place out.

Thus the scene was set for the events that took place on a bitterly cold night in February, 1966. I had gone early to bed, suffering from a recurrence of the malaria that still plagues me at intervals and is a legacy, one I could well do without, of the months I had spent on the coastal belt of Kenya many years before. I slept badly, racked by the lunatic and semi-delirious dreams that are so much a feature of this unpleasant condition. Amongst these dreams was one that was especially vivid in which Random was battling with some gigantic opponent who was throttling her with one hand, apparently indifferent to the terrible punishment he was receiving from her raking talons, which were ripping red furrows in any part of him that they could reach.

I woke some time in the small hours soaked with perspiration, as weak and flabby as a plate of cold porridge, but with the fever mercifully departed. As I lay there thinking of the dream, which was as horrible as ever, I thought I heard the sound of banging and distant shouting, but, weak as I was, I could not pull myself together. The wind was singing and humming through the telegraph wires and from somewhere came the crash of a falling slate. With no feeling of presentiment I finally drifted off into a deep refreshing sleep from which I was awakened by the light in my bedroom glaring at me and by my assistant, George, shaking my shoulder. 'Wake up!' he shouted. 'The garage has been broken into and Random has gone!'

Random stolen? Unthinkable! To me Random was as indestructible and as irremovable as Big Ben or Windsor Castle, part of the fibre of my existence. A sickening chill, icier than the clammy cold that had replaced the malarial fever, swept over me. I thought of those golden eagles that had so recently been stolen from a zoological collection in Derbyshire and had been seen no more. I groped for trousers and a pullover, lurched down the stairs and shuffled into the street. A knot of people was clustered round the open door of the garage. They, it appeared, had heard the nocturnal hubbub but, like me, had ignored it.

It was easy to see what had happened. A broken bottle lay in the gutter. The stout padlock had been forced from its supports. The door was only slightly open and I pushed inside, looking wildly round the dim interior. Random had gone, there was no doubt about that. The inside of the garage was chaotic. The heavy packing-case was overturned. The perch was drooping drunkenly from one corner. I opened the door wider to let in as much light as the grim February morning allowed. In one corner lay a shattered electric torch. At my feet was a piece of material, a bit of mackintosh, perhaps three inches square. It was soaked in blood. Nearby lay another piece of cloth about the same size, obviously part of a tweed jacket. It too was blood-stained. My spirits leapt. Ye gods – Random must have put up a fight and a half, and in almost total darkness too!

With the evidence in my hand I gazed around me. The crowd looked at me and at the bits of material and still said nothing. I put the torch, or what was left of it, in a corner, pocketed the bloody relics and, calling Random's name with little real hope, set out down the road. All trace of my malaise had for the moment gone. I could not have walked more than fifty yards before I heard her calling. I spun round in answer to her summons and saw her, huge and grim in the grey light, perched near the top of an apple tree in the garden of one of my neighbours. Without knocking at the door I opened the garden gate and walked up to the tree where she sat. She was dishevelled, her breast feathers all awry, the tips of at least two of her primary feathers snapped off short, and she was in a filthy temper, but she was alive and she was here. Her raised hackles slowly subsided, the savage light died from her amber eyes. She was obviously pleased to see me.

I scrambled up and without thinking hoisted her on to my ungloved fist and took her home. The knot of people outside the garage had increased in number but they still seemed too awed to say much. I put her on her block, closed the garage door, and rang the police. The local constabulary were both intrigued and amused. Never had such a case occurred before, nor was it likely to occur again. 'I'm afraid, sir, I will have to take the material to use in evidence if an arrest should take place,' said the local bobby in the best Agatha Christie tradition as he stood with poised notebook and pencil. Personally, I was not particularly interested now as to whether or not the miscreant was brought to justice. I doubted very much whether he was likely to try that sort of thing again.

The press, in the person of the local representative of the *Portsmouth Evening News*, followed hot on the heels of the police. He did Random proud. That evening his paper bore the headline: 'Emsworth Eagle Puts Gang to Flight'. This was overdoing it a bit, but no one came forward to dispute the fact. From then onwards most of the national dailies were on the phone. The result next morning was spectacular. 'Random Gives Burglar the Bird' shouted the *Daily Mirror*. The *Daily Telegraph* was more reserved. 'Golden Eagle Swoops on Intruder' was its offering. The *Daily Sketch* gave a blow by blow description of the encounter, which, considering there were only two witnesses, one of whom was only too glad to forget the whole thing and the other for obvious reasons had no intention of saying anything, was a masterly example of what a bit of imagination can do. Random's fame even reached the continent. '*Random der Wachtadler*' said a well-known German paper and the French *Paris Soir* showed a cartoon of an heroic and vengeful Random flying off with a cloth-capped, masked and jemmied burglar dangling despondently from her talons. Biddy Baxter, the producer of the BBC TV children's programme 'Blue Peter' rang me up and I took Random (and the piece of mackintosh which I retrieved from the police) up to London for an interview. Altogether, life was one long Roman triumph for the next few days, until something else turned up to knock us off the world's stage!

Chapter Seventeen

One of the results of all this was that I was confirmed in my belief that I enjoyed the grocery business as little as I had the pet-shop trade. What I wanted was to find some real country, if there was any left, where I could settle down and get on with the business of living. I would like to find a cottage, as remote as possible, where I could live with my few birds of prey and perhaps two or three dogs and follow my real interest – natural history in all its aspects. Emsworth was all right, at least it was better than London, but, as far as I was concerned, only just.

Just after the cuffuffle over Random and the burglar I was asked to give a short talk on the radio (the wireless, as I still thought of it, and still do). I talked about Random, about eagles in general, and finally about Cressida the kestrel. A few days later I got a letter with the heading 'Kingswood School, Bath'. It was written by my cousin Peter, who was, and had been for many years, bursar of that long established seat of learning. He invited me to bring Random to the school and to give an informal talk to the natural history society. Enclosed in his letter was a photograph of Cressida and of my old lurcher bitch, the first Bracken, which he had cut out of *The Listener* together with extracts from a talk I had given on Children's Hour in 1947.

I took the train from Emsworth to Bath with Random concealed in a huge coffin-like travelling-box, causing much speculation amongst fellow travellers. This box, about four feet long and three feet high, was the source of much embarrassment. It seemed virtually impossible to put it anywhere without people queueing up in serried ranks for what appeared to be the sole purpose of tripping over it, which resulted in filthy looks and uncharitable language from a surprising variety of what were normally, no doubt, the kindest and most tolerant of citizens. It was much too big to go into the rack, it took up virtually the whole space between the seats, and the only solution was to park it in the corridor and to gaze aloofly out of the window in the hope that the enraged passers-by would believe it to be the property of the innocent-looking bloke sitting opposite me reading the *Financial Times*.

My cousin Peter, who met me at Bath station, accompanied by a posse of his natural historically-minded pupils who had been dragooned into acting as amateur porters, was unlike anything that I had expected. Tall, lean and neatly bearded, he appeared to me like a sixteenth-century Spanish grandee. I could picture him clad in doublet and hose and wearing a big plumed hat, briefing the crews of the Great Armada, before that luckless fleet left for the Channel. Random, lodged in the natural history hut, made her presence felt by decapitating an entire and probably irreplaceable collection of stuffed birds, doubtless the life work of some long dead Victorian or Edwardian taxidermist. I hope he would have appreciated the irony of having his beautifully mounted specimens reduced to headless ruins by a live and exceedingly unappreciative golden eagle, a bird on which he would no doubt have given much to lay his hands sixty or more years previously.

The talks were, I believe, a success and proved useful training for what was to follow in later years. If you can tackle an audience of schoolboys you can take on anything, and, judging from the silence during the lecture and the questions asked afterwards, not to mention the awed admiration for Random which was evident from her first appearance, the journey, with all its discomfort and embarrassments, was well worth while. Over a glass of Madeira later that evening Peter asked me casually if I had ever considered writing the story of Cressida's adventures. I had in fact thought of it often but dismissed it as being well beyond my capabilities. A few short articles in sporting magazines were all that I had managed by way of literary output up to that time. True, I had once, whilst waiting to leave Britain for Africa not long after the war, written a few pages about my early associations with Cressida but I had torn them up in disgust. The whole thing looked trite, over-sentimental and thoroughly badly written. The literary life was obviously not for me.

Peter, it seemed, was certain that the time had come to try again. According to him there were not enough animal stories or tales of adventure to go round. The Cressida saga was a mixture of both. Therefore, according to him, it couldn't fail to hit the right target. I agreed to all this in principle, told him I would try again some time, and promptly forgot the whole thing. I enjoyed my visit to Kingswood. It brought back to me a world I had almost believed had ceased to exist. I hadn't been near a school since 1937 when I had left Gordonstoun and I had no particular wish to repeat the experience. Now it was very different. I couldn't help being rather sardonically amused as I thought of those days thirty and more years ago when I had regarded lecturers with awed veneration. Now, here I was receiving much the same sort of respect. I remembered the thrill of admiration and envy with which, in the early thirties, I had regarded the redoubtable Captain Knight and his golden eagle 'Mr Ramshaw' and how I had become the temporary hero of my friends and enemies when I had actually been allowed to hold the colossal eagle on my fist for a few tremulous seconds. Here I was with an eagle of my own clutching my sagging fist and staring round imperiously at the enchanted audience.

On my return home I put the grocery business in the hands of the nearest estate agents. The fact that this brought forth no immediate reaction surprised me not at all. Every second shop in the village appeared to be offering some form of foodstuffs in addition to what they were supposed officially to specialize in. Furthermore there were at least three bona fide grocery stores standing cheek by jowl within a few yards of my own establishment. Despite this ever-increasing competition I managed to take enough money to support my animals, which after all was what the operation was all about.

None the less I had come to the grim conclusion that I had more birds of prey than I could justifiably maintain. It wasn't just a question of giving them enough food; it was equally important that they should have regular exercise in natural conditions and I had not got the time or the space to fly them all as consistently as they should have been flown.

Razmak was the chief problem. She was as active, restless and sensitive as a goshawk. For her size she was exceedingly predatory and was obviously designed by nature to prey on rabbits and other small ground game, which didn't exist in that part of the country in sufficient numbers. Myxomatosis was still reaping its disgusting harvest, and brown hares, which would have suited her admirably, were thin on the ground, being found almost exclusively on the big local estates, whose owners, being keen shooting men, showed a regrettable, if understandable, reluctance to allow so lethal-looking a bird as a hawk eagle to fly unhindered about their pheasant-infested properties.

Bianca, the black collared hawk, was quite different. Being a good deal tamer than and about as innocuous as the average domestic fowl, she presented few problems. She showed little inclination to catch anything larger than a half-grown rat, and that only when she was exceedingly sharp-set. She was fond of her home and, provided Random was out of the way, could be allowed her freedom. She was as unlikely to take her departure or to get into trouble as would be a fan-tailed pigeon.

George Mussared had returned from his cruise in the *Eagle*; had, in fact, just left the navy, and had settled down in his native Yorkshire. Knowing that he would give Razmak the sort of home that I would like to have given her myself, I wrote to him, mentioning my problem. George lived within easy reach of the open Yorkshire countryside, with its moors and wolds still teeming with game, and I knew well enough that he would see to it that Raz lived as natural a life as was possible for a trained hawk, an alien in a strange country, to enjoy. Within twenty-four hours he had presented himself at my door to collect her and within a week was flying her loose at natural quarry. A partnership had begun that was to last for a number of years. A few days later I gave Bianca a freshly killed blackbird that I had found by the side of the road. Believing it to have been hit by a car, as indeed it had been, I hardly gave it a thought, and Bianca ate it with gusto. The same night she died of insecticide poisoning.

Once more Random and I were alone together, with the exception, of course, of Bambi and Bracken, who were still very much part of my life. The fact that there were no other birds of prey to come between us suited Random admirably. Had she had the opportunity there is little doubt that she would have seen to it that such a state of affairs had occurred long before. Over the years she had become increasingly possessive and jealous over our relationship and would attack without hesitation anyone whom she thought might come between us. She would also go for anyone, stranger or otherwise, who intruded upon the piece of ground that she regarded as her own territory.

She hated some more than others, and there was one particular man whom she detested with an almost maniacal fervour, a hatred which was terrifying to witness. On hearing his voice, even at a considerable distance, she would make frantic efforts to get at him, shouting abuse in her ringing, far-carrying voice. She would leap up and down with half-extended wings, her hackles raised and with flashing eyes. I was haunted by the thought that one day he would appear while she was free, an encounter that might well have disastrous consequences, and I thanked providence that I was not on the receiving end of this single-minded fury. In this particular case she showed shrewd judgement, because the fellow proved to be a particularly unpleasant individual and was eventually successfully prosecuted for barbaric cruelty to a number of tawny owls which he had caught in pole traps. Had I known about this at the time I might well have taken less care to secure her on his frequent and none-too-welcome visits.

For a short time I employed an elderly villager to help keep the premises tidy and do odd jobs about the place. He made no secret of the fact that he disliked and feared Random, a dislike which she returned in full. One morning I found him, for no reason other than natural spite, threatening her with a broom. She showed little sign of reaction other than that her hackles were lifted slightly and that there was a hard, calculating look in her eyes. She was still perched, apparently relaxed, on her block with the leash coiled loosely at her feet. None the less, I knew that the man had made a dangerous enemy and I told him so in a few brief phrases. A few days later I saw him waving the broom at her again and cursed him roundly.

Perhaps a fortnight had passed before, during a period of exceedingly high winds, I happened to be in the shop when an appalling din arose from the kitchen. I hurried out to find the old man, or most of him, in the sitting-room with the door partially closed. One of his feet, with Random firmly adhering to it, was still in the kitchen, while the ululating calls which arose might well have been emitted by the Mahdi's hordes before the battle of Omdurman. In short, the old boy did not appear to be enjoying the experience unduly.

I grabbed Random round the body and she let go, after giving one or two half playful squeezes to show that she still bore plenty of ill will. I hurried her to her block and returned to inspect the damage. Luckily for him the victim, as was his wont, was wearing a pair of tough, old-fashioned boots which reached well up above the calf and so, apart from a bit of bruising, no harm had been done. He had learned one lesson — never, if you can avoid it, make an enemy of a golden eagle. It appeared that the wind had blown the garage door down during the night. Random had emerged at dawn and had taken her usual stand upon the block, where she no doubt rested in her dreamy way, contemplating the infinite, until the arrival of old Mark reminded her that she had a score to settle. Flying the length of the garden, passing through the lean-to, the scullery and the kitchen, she rapidly went about the business of settling it. Thereafter she entirely ignored his presence and he for his part gave her an exceedingly wide berth — a state of affairs that suited me admirably.

An acquaintance of mine who was at the time serving in the RAF at Thorney Island owned one of the fastest salukis in the country, a smooth-coated fawn dog called Amir. We had been out several times together to Harting Down and on one notable occasion had been nearly laid low by a cossack-like charge from a troop of fallow deer, who, with Tally and Amir at their heels, had almost overwhelmed us with the speed of their coming, as, unaware of our presence, they swept down upon us in their determination to reach the beckoning thickets behind us. This friend of mine lived near Llandovery, in the county of Carmarthenshire in South Wales. He frequently invited me to stay with him, an invitation which I as frequently turned down. To me in my abysmal ignorance Wales at that time was much like Outer Mongolia, a place to read about but under no circumstances to visit. I believed it to consist of interminable coal-mines in the south which gave way, somewhere to the north of Aberystwyth, to equally interminable slate quarries.

I was aware that there must be a considerable area of wild, mountainous and wooded country, because I knew well enough that the last pitiably small remnant of a once flourishing and widespread species of raptor, the red kite, had its British stronghold somewhere amongst those wooded valleys, but whilst I was pleased to know about this I was, after my wanderings abroad, a bit blasé about kites, both red and black (the latter being in fact little darker than the average buzzard). In East Africa kites were everywhere, drifting lethargically overhead or perched squealing petulantly on every suitable vantage point, waiting for anything small, edible and of animal origin, dead or alive, that might appear and stir them into the wheeling, diving, noisy activity that one associates with herring gulls at a British seaside resort. I also knew and accepted without undue enthusiasm the fact that the Principality was the home of ravens, pole cats, pine martens and other fabled beasts which somehow I had always associated with the highlands of Scotland.

Eventually, after my third or fourth refusal and believing that any experience was worth while, I accepted the final invitation, partly because I realized that I had been doing that lovely country an injustice and partly because I was fed up to the back teeth with the flat, featureless and over-populated southern coastline and thought that any change would be for the better. In any case I would be away for less than a week. On a Sunday evening, late in June, 1967, with Random in her travelling-box and the three dogs (I had recently acquired an Italian greyhound bitch puppy named Kiki) peering expectantly out of the windows of my friend's car, we pulled up at the approach to the Severn Bridge. As I gazed down at the grey waters of the Bristol Channel far below, little did I realize that I was about to cross my Rubicon. From that day forth life was going to take on a different shape. We paid two shillings at what looked excitingly like a customs barrier and, feeling vaguely disappointed that I had not been asked for my passport, we shot forward, crossed the bridge, and entered another world.

Chapter Eighteen

Much has been and doubtless more will be written about Wales and the Welsh by others far more qualified to do this than myself. Until the day I crossed the Severn Bridge I had always imagined, if I thought about it all, that entering the country would be much like moving, say, from Worcestershire into Gloucestershire. It was not in the least like that. The impact of a new country with new people and a different way of life started on arrival at Chepstow in Monmouthshire and increased with every mile as one penetrated into the interior. Wales seemed then, as it seems now, to be unique in all respects, in its topography, its people and above all in the atmosphere which is so much easier to experience than to describe.

The relationship between Wales and England seems to have something in common with that between Luxembourg and France or between Austria and Germany. This similarity, however, is no more than superficial. The fact that a relatively tiny country, a country bordered on its eastern side by one of the most densely populated countries in the world, should proudly maintain its own language and culture commands much respect. This is the language that must have been spoken around the camp fires by the truculent hosts of Boadicea, a language that can have changed but little during the passage of the centuries.

I have in my possession a bird-book written during the mid-nineteenth century. This book, by the Rev. F. O. Morris, gives not only the English and Latin names of the birds described but also the names by which they were known to the Ancient British. For instance, in the case of the great grey shrike, the name was *Cigydd Mawr*, which in modern Welsh means, of course, 'the great butcher bird.' Surely this suggests that the language used by the inhabitants of the Principality today is the same as that spoken by the men who lined the cliffs of Dover to welcome the first wave of Roman invaders? Not even the most ardent Anglophile could say that the English language has remained as uncontaminated by outside influences throughout the ages!

The people, furthermore, are quite different from their neighbours to the east of Offa's Dyke. In fact, I would state categorically that the crowds which throng the streets of some small town in north-west France look more 'English' than would a gathering of farmers grouped around the cattle pens at Llandovery mart. The sturdy, white-washed farmsteads dotted about the gentle, grey-green hillsides have an air of permanence, of laughing at time, that gives the impression that come what may, whatever horrors engulf the outside world, those who live in them and farm the changeless uplands on which they stand will be there, immovable, weathering the worst that fate can throw at them as they have weathered the past few hundred years.

Although now by force of circumstances the majority of them live in towns the Welsh are at heart country dwellers, and mountain dwellers at that. Like all those who live close to mountains they have acquired a cheerful philosophy, a love of music, laughter and good fellowship, and above all a deep and abiding passion for their homeland that can best be described in the one word *Hiraeth*. This implies a blend of yearning nostalgia, heartfelt pride and sense of unity which is particularly evident as soon as one starts talking to an expatriate Welshman. I remember it as the outstanding characteristic of those I met in Kenya and other outposts of what was then the British Empire, yet I never fully appreciated it until I met the Welsh in their own environment. It is, in a rather different form, found also in the Scots, and, having a good deal of Scottish blood myself, although regrettably diluted, I can feel an innate sympathy for this intense, deeply imbued patriotism.

One more fact that I consider worthy of interest and respect is that the Welsh appear to have remained largely uninfluenced by successive waves of invaders and their blood is probably as pure, or almost so, as it was before the Romans arrived. They do not appear to have intermarried to any noticeable extent and as such have strongly characteristic racial attributes. The women, by and large, are prettier and more attractive than their neighbours and have a natural vivacity which is almost Mediterranean in its spontaneous cheerfulness. All in all, the Welsh are an interesting and admirable addition to the British fauna!

Somewhere between Brecon and Sennybridge I saw the first buzzard I had seen for years. It was sitting unconcernedly and stolidly on top of a telegraph pole. I was so excited that my host stopped the car so that I could examine the bird in detail. It was a large female with a creamy breast and throat — buzzards, I soon discovered, could be almost anything in colour from practically albino to almost completely melanistic. The French name, *Buse Variable*, describes them admirably. This one took off like a great, lazy, heavy-bodied moth, curving out over the fields to swing back and alight upon another pole a hundred yards further on. Within a mile or so I had seen two more buzzards, both differently patterned, one having noticeable horizontal striping on its breast that was almost reminiscent of a goshawk. A day or two later I became so accustomed to seeing them that I hardly bothered to look up when I heard the half-wistful, half-petulant mewing call overhead and knew that the wide-winged, almost aquiline shape was sweeping in lazy spirals above me.

I was, and still am, amazed at how deceptive their size can be, their four-foot wing-span appearing at times and under certain conditions of light and distance to reach a good six feet or more. Little wonder that they are on occasion mistaken for eagles. Buzzards are unexpectedly common in Carmarthenshire and Breconshire and, being born opportunists, have quickly learned to take advantage of the havoc wreaked by motor-vehicles. Bone idle by nature, they will wait for hours for something to commit suicide, whereupon, with no effort to themselves, they will drop down and feast copiously. I was to find later that a tame buzzard can be exceedingly unpredictable in temperament and will, if in the mood, attack its owner or go for larger birds of prey with an almost frightening determination and savagery.

My friend, Neville Evans, was living at that time in a farm a few miles from and above the hamlet of Babel. I say 'above' because the house and its satellite buildings stood on a piece of ground more or less chiselled out of the steep, grassy and tree-dotted hillside behind. The whole place, and indeed the whole area, reminded me of the White Highlands around Thomson's Falls in Kenya. There were the same deep red-soil valleys with their thick, mossy trees canopied in apparently perpetual mist and below these the swift-flowing, rocky streams, haunted by the dipper and the pole-cat. I almost found myself peering aloft for the troops of black and white colobus monkeys which are so much a feature of the African forested uplands.

One thing these two different yet strangely similar landscapes had in common was the number of buzzards. Of course the dramatic-looking pied augers with their chestnut-red tails and loud clanging cries of 'aung-aung-aung' were replaced by the drabber, rather smaller European buzzards but the general impression was much the same. There was, however, one hazard that I had not encountered in East Africa – sheep. The whole countryside was packed with them and they hung around the outer defences of the farm like a besieging army. I knew that Bracken at least, a huntress to the marrow of her bones, would, if given the slightest opportunity, take off after them and thereby cause a diplomatic incident and probably end up by being shot for good measure. This was to prove a persistent nightmare. I had known vaguely that Wales was a sheep-farm-

ing country but it had never occurred to me just what this implied. However, this was a problem that had to be overcome, and overcome it I would – otherwise, my short visit would be so many days of unadulterated hell.

Random, tired and stiff-winged from her long journey, settled down on a prefabricated block, a large upturned packing-case, in a clean, airy and spacious barn. I was relieved to notice that, weary as she might be, she was not too exhausted to make an exceedingly square meal out of a fresh sheep's head which she rapidly reduced to splintery remnants of hair and bone. After this, her crop the diameter of a medium-sized football, she drew up one leg into her breast feathers and fell into a satisfyingly full-fed reverie. My three dogs, after a certain amount of hackle-raising, stiff-legged formality, settled down with Neville's half-dozen assorted working terriers and all slept together happily enough in the thickly straw-strewn kennels close to the house. I was a little bit uncertain at first, as none of my brood had up till then slept anywhere except on mats or chairs before a blazing fire, but I needn't have worried. After a few experimental, half-protesting whimpers, they all curled up together and slept, tired out, in the warm, rustling darkness.

I slept well that night with the pleasantly exhilarating feeling of being in foreign parts, a feeling that was reinforced to some tune when I was jerked awake by an unearthly din – a medley of shouts and raucous calls in an unintelligible language that for a few seconds made me feel that I had fallen into the clutches of a gang of Corsican bandits. In my sleep-stupefied brain I tried to work out what was happening and where I was. Peering upwards in the grey light of dawn I saw what I hadn't noticed when I went to bed – a large square skylight. Staring through it were the black heads of what in my bemused state looked not so much like Corsican bandits as a group of cannibals from the Solomon Islands examining the prospective menu. When my eyes managed to focus properly I was relieved to discover that they were nothing more lethal than a gang of carrion crows; crows, however, with the most extraordinary and wide-ranging vocabulary. Neville told me later that these crows came regularly to steal lumps of putty which were supposed to keep the skylight in place. No doubt, on finding the room occupied, they had been unable to control their surprise and indignation.

I had always believed, and indeed my own observations had

done much to confirm my belief, that carrion crows were anti-social by nature, enjoying their own company and that of their mates while avoiding those great social gatherings which are such a notable characteristic of their relatives, the rooks. I had always considered that their vocal accomplishments were limited to the snarling, nasal 'karr' thrice repeated, which seems to be the signature tune of this unpopular and un-scrupulous bandit, a bird whose outward appearance in no way belies the evil of its ways and which has been condemned throughout the ages as a merciless murderer of young and helpless things. This may well be true but there is another side to the crow's nature that I was only to discover when I re-luctantly accepted the role of foster parent to a whole series of fledgling crows. My conclusion was that the carrion crow is about the most intelligent and adaptable of all birds. The fact that it has learned to seek sanctuary in the heart of our largest cities, where it knows perfectly well that it will be tolerated if not actually welcomed, is proof enough of its pioneering spirit.

Cursing this clamorous and uncalled-for deputation, who were still peering down at me and quite obviously making un-charitable remarks about my personal appearance, I dressed and went down the twisting, creaking staircase to the kitchen and out into the farmyard. I peered through the cob-webbed window of the barn. Random was already awake and doing her 'flying on the spot' exercises, no doubt with the idea of relieving the stiffness in her wings. In the centre of an empty paddock surrounded by wire-netting arose the stump (per-haps three feet high and as level as a table top) of a recently felled tree. A hasty search produced a box of rusty but other-wise serviceable staples and although a hammer was not forthcoming I managed to drive one of the largest of the staples into the side of the stump with a handy, flatsided piece of rock, leaving enough space for Random's leash to pass through.

I carried her carefully into the paddock and with a forward movement of my arm encouraged her to leap on to the top of this ideally situated natural block. I fastened the leash with a falconer's knot to the staple and stood back to watch her as, with wings and tail extended, she revelled in the first rays of the sun, which was just clearing the gently undulating hilltops to the east, bathing with a pearly radiance the dewdrops on the meadow grass, the streamlet chuckling down the slope and the pink and white roses in the hedgerows. Random looked about her, roused her feathers, and cocked her head to gaze sky-wards at the pair of buzzards drifting high overhead. She looked as much at home, as completely a part of this wild landscape, as if she had lived there all her life.

On impulse I untied the leash, lifted her on to my arm, and climbed the slope above the farmstead. It was a steep climb and Random was heavy, but I went upwards until we were high above the grey slate roof-tops, looking out over the sheep-sprinkled pastures which dropped away below us to the valley floor beyond which, purple in the early light, arose the distant outposts of Mynydd Eppynt. A deep, challenging triple bark made me look upwards. High overhead a pair of great black birds with wedge-shaped tails and forward-jutting heads from which protruded pick-axe beaks, were driving across the sky. High as they were I could hear the rhythmic swoosh of their wings as they surged forward with a contemptuous ease that no mere crow could emulate. I had never seen a wild raven but I knew them at once, before even the full-throated croak, thrice repeated, came down to confirm my identifica-tion. One of the birds, it might almost have been for my benefit, closed its wings and rolled earthwards in a corkscrew dive, whilst its mate promptly followed suit, apparently for the sheer hell of it. Gaining height once more the two birds vanished into the haze with another volley of derisive croaks which held in them something of the spirit of the Welsh hill country. To me, from that day forward, the ravens' craggy croak has always been the real voice of Wales, capturing the essence of freedom and defiance that is so much a feature of that country and its splendid people.

198

Somewhat light-headed with the sense of elation that had overtaken me and with little thought for the consequences I slipped Random's leash from her jesses and held her aloft. With an upward lift of her wings and a powerful back kick of her great yellow feet, she was away. To watch a golden eagle sweeping above the gentle domesticated fields of the English south country is impressive enough, but to see the same bird in action against the backdrop of the Welsh mountains is another thing altogether. Random seemed to double in stature, to take on an air of controlled majesty that even I, who had known her in all her moods, had never fully realized she possessed.

Perhaps my perception was heightened because it occurred to me at that moment that this possibly was to be the parting of our ways. If it was to be so, I wouldn't have had it otherwise. The choice was hers and hers alone. We had been together for nine years, not all that long perhaps, as one judges time, but it seemed an eternity to me at least. She had come to be an integral part of my life. What she thought of me is pure conjecture but judging from the way she behaved, her greeting of me after my absences, and the fact that she had always returned to me of her own free will, I was confident that, despite her limited power of expression, the love and respect I felt for her was returned in full. This belief in her affection and feeling of comradeship was now to be tested.

As she slanted out, swinging far beneath the shadow of the hillside before climbing into the upper air, the sun caught and gilded the metallic feathers on her nape and hackles, turning them to richest copper. Her primaries caught and fingered the rising air currents, toying delicately with the gentle breeze. Half closing her wings she hurtled earthwards; then, checking her fall, once more she swung out over the valley which dropped away almost vertically below her. For the first time I had the experience of looking down on her as she cruised in wide circles, glorying in the freedom of her natural environment.

Her telescopic eyes must have noted the distant forest below, with its hint of wild game to be chased and seized at will. The pair of buzzards towering on high saw this immense bird, a bird the like of which they had never seen before; only the occasional heron, laboriously flapping in a straight line from fishing ground to heronry, had anything approaching this interloper's width of wing-span, and the buzzards knew the herons well, knew that they were inoffensive anglers offering no threat. This was a different proposition: a bird like them in form and outline but with nearly double the wing-span and five times the weight, a bird that by its very presence offered a nameless menace.

The big broad-winged hawks' plaintive mewing call took on a sudden hint of savagery, as, with the larger female leading, they closed their wings and fell out of the sky with a hissing stoop of which even a peregrine might not have been ashamed. Whether the indignant hawks intended to press home the attack is impossible to say. Despite its humble diet of rodents and carrion the buzzard is no coward. Even the raven, its equal in size and its superior in armament, although ever ready to mob the deceptively leisurely-looking raptors, is wary of getting within reach of the buzzard's short, stubby, but exceedingly powerful talons, preferring, by taking advantage of its masterly wing-power, to chivy its slower rival until in sheer exasperation it takes refuge in a tree. Yet despite this apparent unwillingness to engage in close combat, I have noticed that, if a buzzard is intent on going somewhere specific, it goes there, irrespective of the gang of noisy black hooligans at its tail.

Random was enjoying herself, giving herself up to the sheer pleasure of the moment as she breasted the broad bosom of the strengthening wind which bore her up, playing with her as if she were a piece of drifting gossamer. Listening to the air humming through her slotted primaries she did not at first hear the rushing descent of the approaching foe. Something, instinct perhaps, made her glance upwards. The female buzzard was only about fifty feet above her and was coming down like a six-inch shell, closely followed by the male. Random, never slow to accept a challenge, had no intention of shirking this one. Apparently standing on her tail, she went up vertically to meet the enemy head on. The buzzards must have realized in that instant that here was an opponent such as they had never encountered before, but with their weight and the speed of their descent they were unable to pull out of their headlong dive. Just when a collision seemed inevitable and Random, nimble as a matador, had thrown herself on her back to take the brunt of the attack, some lightning reflex action came into play. As a rushing mountain torrent divides and by-passes a boulder in mid-stream, so the two buzzards by-passed Random; even as her clutching feet shot out to grapple they were safely past and continuing earthwards with undiminished speed. As the grassy hillside rushed up to meet them they levelled out and swept into the branches of a solitary oak.

The whole performance had taken only seconds but it was a magnificent display of natural artistry faultlessly carried out by all three participants. Random righted herself and, once more on an even keel, looked round to see what had become of the opposition. Finding herself alone in the sky she banked sharply, waggling her wings in what looked to me very much like a modified version of the victory roll, and slanted downwards in one long gentle gliding movement. She landed within a few feet of where I was standing enthralled. Random had proved once again that, despite all the obvious attractions of this intriguing new country, with all its hazards and challenges, she still, at least for the present, preferred to throw in her lot with me.

Chapter Nineteen

I carried Random back to the paddock and left her on her block whilst I went to fetch a special treat. I had brought with me from Emsworth a fresh kidney, which I knew was one of her favourite foods, especially to celebrate her safe return (as I hoped) from her first free flight in this new landscape. I gave it to her and watched as she held it down with her talons, tore it into small pieces and ate it with infectious enjoyment. I sympathized with her. There are few things I enjoy more, though not, I confess, in its natural state. Having eaten the last gory particle she absent-mindedly wiped her messy, scarlet-flecked beak on the stump, drew up one leg and sank into a state of dreamy reminiscence. I realized that she wished to be alone with her thoughts. I believe that it is partly this respect for her occasional wish for privacy that has built up the feeling of mutual esteem that has made our relationship so rewarding. I also feel the need for moments of complete solitude, so I can understand just how she feels.

I returned to the farmyard and opened the door of the kennel. The dogs poured out in a multi-coloured river of movement – black and white, fawn, blue, and tri-colour – like a pack of miniature fox-hounds. They had slept well, were warm and comfortable, and ready for anything that fate might have to offer. The terriers, short-legged, stump-tailed, and of all types of coat, rough, smooth and 'broken', cavorted and tore round the yard, pretending to look for rats in every likely or impossible situation, all with the curious terrier characteristic of favouring one hind leg as if pretending to be lame. After a suitable display of hunting fervour, in which they did their best to justify the sort of advertisement one so often sees in the livestock papers – 'Jack Russell Terriers. Will Face Anything. Dead Game to Fox and Badger' – they gathered by the gate at the top of the yard, as if knowing exactly what to expect. I felt, contemplating this pack of jaunty, raffish, little tykes, that they would have been ready to face rhinoceros without the slightest misgiving. I was relieved to see that they had apparently accepted my trio of interlopers with unexpectedly good grace, which was as well, because they looked as if they would make much better friends than enemies.

I was particularly taken with one of them, a little white bitch with a beautifully 'badger pied' head, who rejoiced for some reason known only to her owner in the lamentable name of 'Meece'. This interest was reciprocated, as from then on she more or less adopted me and seldom let me out of her sight, to the intense and vociferous indignation of Bambi and Kiki. Bracken, silent and aloof as her saluki sire, said nothing but made it clear enough that she would not tolerate any attempt at stepping out of line on the part of this uppish and impertinent terrier.

The farmyard was luckily entirely surrounded by sheep-proof and almost dog-proof (or at least large dog-proof) fencing. To get out and into the pasture land beyond I had to climb a steep path and open or climb over a rusty iron gate (which was fastened by an almost incomprehensible tangle of wire and string). Neville had told me that his dogs were steady with farm stock but I had no idea how mine would react, although I feared the worst. In order to avoid the expected slaughter I fastened a lead to Bracken's collar and, after an excruciating battle with the gate and its fastenings in which my hands and temper were both equally lacerated, I managed somehow to force my way through, followed and preceded by my disreputable throng, all yelling their heads off at the prospect of the chase. Fortunately the first wave of sheep, alarmed by the din, had retreated to higher ground, where they watched our approach with increasing disapproval.

I turned left along a hummocky track, liberally sprinkled

with fresh black sheep-droppings, evidence of the number and ubiquity of the enemy. On one side rose a bank covered with wiry-stemmed vegetation, bramble, briers and blackthorn, richly spangled with pink and white trumpets of convolvulus. This bank was riddled with ancient rabbit burrows whose tenants long before had wisely moved elsewhere. Into these cobwebby labyrinths the entire gang of terriers promptly disappeared, leaving my three, who had never seen such a performance, gazing at the blank, leaf-strewn holes with amazement. The terriers knew as well as I did that the burrows had long been unoccupied but were quite obviously putting on a show for the benefit of their visitors. Bambi and Bracken, who knew something about rabbits and rabbiting, bounded from hole to hole in the forlorn hope that something, anything, might emerge. The only things that did eventually surface were the terriers themselves, who issued fourth from the hidden depths like so many Orpheuses (Orphei?) from the Underworld, caked in mud, their erstwhile shiny black noses red with clay and their fore-paws so encased in the sticky soil that they looked almost as if they were wearing small glutinous boxing-gloves. They were, however, happy and laughing-eyed at their own daring in penetrating deep into the dark, mysterious galleries beneath my feet.

With their initial impetus spent they were a bit more controllable as we followed the winding track which led enticingly onwards towards the fringe of dark, primeval-looking forest ahead. I had released Bracken, who padded behind me, her nose within a yard of my left thigh, apparently as disciplined as an experienced gun-dog, and was beginning to feel pleasantly relaxed and at one with the world about me. Then, rounding one of the innumerable bends, the pack nearly collided with a trio of speckle-faced mountain sheep, led by an enormous ram, horned and bossed like a buffalo bull. The terriers stood stock-still, only their stumps twiddling uncertainly, their ears half pricked with expectancy. The two Italian greyhounds, who had never encountered such beasts before, rushed forward shrieking hysterically, and the sheep turned and crashed, as sheep will, straight into the entangling clutches of the thicket.

Acting purely by instinct I stooped and seized a clod of earth, which I hurled at Bambi, bellowing with rage and fear. At that moment, for aught I knew, jealous agricultural eyes, peering from some unseen vantage point, might well be focused upon us, watching our every movement, fingers poised over the safety catch of the twelve-bore, ready to wreak instant vengeance on anything that threatened the safety of these mobile lumps of mutton. Luckily my aim was true and the clod caught Bambi amidships, silencing him in mid-yap. All the dogs, now innocence itself, pressed round me as I travelled onwards. The sheep had forced their way through the hedge on to the open range beyond and were grazing as if nothing had disturbed the even tenor of their ways. Bracken, I noted with satisfaction, had paid practically no attention to their existence, or so it seemed at the time.

So distraught was I at this hair-raising incident that I failed to notice that the path had finally lost itself amongst the isolated trees that stood like sentinels at the entrance to the forest that loomed dark and enticing as those African rain-forests that I remembered so well. All was silent save for the scampering feet and quick, panting breath of the dogs, who, unimpressed by the sombre beauty of the woods, were ranging everywhere, investigating each hollow, each likely-looking cavity amongst the mossy tree-trunks, ever alert for the scent and sound of some furtive quarry that would unleash their pent-up hunting instinct.

I was struggling along the slippery edge of a ravine, below which a shallow, rock-strewn river chuckled and gurgled, when a sudden, nerve-tingling explosion nearly caused me to lose my footing. It was only a party of scared wood-pigeons crashing out of the summit of a near-by beech tree, but in my tense state it was as alarming as a volley of rifle fire. Looking up, I saw the reason for the pigeons' sudden departure. Swinging between the branches, now heavy with mid-summer foliage, came a large female buzzard. Manœuvring with un-expected grace and dexterity for so bulky a bird, she circled over my head at no great height, yowling like an ill-tempered tom-cat. This savage challenge was echoed by a higher, shriller call from far up in the sky, well above the forest canopy. Her mate was egging her on but himself taking good care to keep out of the range of possible reprisals.

All thought of sheep and shepherds temporarily forgotten, I moved further into the depths of the forest. The buzzard swept close above me, her voice becoming more and more abusive. Intent on keeping my foothold on the precipitous, slippery hillside, made treacherous by the decaying mould of many seasons' fallen leaves, I grasped at the stems of con-venient saplings to pull myself up the slope. The dogs, agile as chamois, scrambled unconcernedly in my wake. The buzzard's agitation had reached a crescendo when, glancing upwards, I saw what at first I took to be the flattened remnants of a squirrel's drey, snugly fitted into the fork of a slender beech tree growing at a curious angle from the hillside, no great distance above the spot where I was floundering upwards. With feet and hands and the help of projecting roots and low-growing bushes I hauled myself vertically up the side of the hill until I was well above the tree that had intrigued me so much. I was now able to look directly down on to it.

The buzzard had taken stand in a spruce tree close by. Her aggression had, it seemed, turned to pleading, her calls now pitiful in her distress. What I had at first mistaken for a drey was in fact a beautifully made nest, surprisingly small for the size of the family it contained. Peering up at me over the rim of the nest, their powder-puff covering of greyish-white down just beginning to disclose their sprouting, deep umber feathers, were two well-grown young buzzards, completely undisturbed despite the consternation of their mother. They were happily nibbling at fresh leaves that seemed to sprout from the actual foundation of the nest. One of them made a

playful snatch with its ridiculous fledgling foot at a bluebottle, obviously attracted by the unlovely remains of a previous meal. They gazed at me fearlessly from their smoky grey eyes and then, losing interest, began to preen the soft blue blood quills of their future tails.

I watched entranced for a few minutes before, feeling that the mother had had more than enough to put up with, I turned and went sliding and slithering down the slope until I reached level ground once more. Eventually I left the forest at almost exactly the same point at which I had entered it. I was so cock-a-hoop at having, within less than twenty-four hours of my arrival in Wales, come across a flourishing family of one of my favourite birds that I forgot entirely about sheep; and the dogs, who had enjoyed the expedition as much in their way as I had, remained close beside me during my uneventful return to the farmhouse.

Some friends of Neville's living near Llandovery had heard through the grapevine (which works as efficiently if as inaccurately in Wales as anywhere else) of Random's arrival in the vicinity and expressed a wish to see her in person. After a late and leisurely breakfast of eggs and bacon, which for some reason tasted far better than the same ingredients would have done in Emsworth, we loaded Random, travelling-box, block and all into the car and set off along the narrow, winding lanes, being nearly massacred *en route* by a Land-Rover driven flat out on the wrong side of the road round a blind corner. This, as I was soon to discover, was by no means an unusual hazard in that part of the world. In fact, it is with relief and amazement that one ever arrives safely at one's destination.

Neville had one appointment to keep on the outskirts of Llandovery, an appointment of which, up to that moment, he had told me nothing. We parked the car and knocked at a door. After a pause, heavy footsteps approached and the door was flung open by what might have been the Lincolnshire Poacher himself, if, of course, this had not been in Carmarthenshire. A brief conversation followed, of which, it being in Welsh, I understood not one word. Our host led us down a narrow path behind his house.

Suddenly the way was barred by the silent arrival of one of the most ferocious-looking dogs I have ever had the misfortune to encounter. About twenty-six inches high at the shoulder, she, for it was a bitch, had the head, pricked ears, and punishing jaws of a bull terrier on a greatly enlarged scale. She was of a rich red mahogany brindle, a real old-fashioned colour such as one seldom comes across these days. It was, however, the look in the eyes that brought me to an immediate halt. Deep set and sullen, they were glowing with a flickering green light, which held a fearful menace, a wordless command to proceed no further. As her jaws were within a few inches of my thinly clad leg I complied with her wishes, hoping that her instinct would lead her to realize what a decent, kindly, dog-loving fellow I was.

The reasons for her hostility now made themselves evident. From the open door of a small shed behind this female Cerberus issued a waddling, tumbling throng of miniature replicas of herself, who hurled themselves upon their mother in a welter of overwhelming affection. At this juncture their owner pushed past and spoke sharply to the bitch, whose whole attitude and expression changed so that she might almost have been another dog. Her wolfish ears flickered and dropped, her tail waved slowly, and her morose expression softened to a look of deepest love and respect. Obeying his spoken command and pointing arm she trotted into the shed, which her owner, to my heartfelt relief, bolted with a satisfying snick.

Some weeks previously, I had rather unwisely mentioned to Neville that I was thinking of getting a lurcher puppy, to add dash and variety to my pack and as a sporting companion and possible mate to my half-bred saluki bitch, Bracken. I have always had a passion for lurchers, which I consider basic, unspoiled, and altogether 'doggy' dogs, as much a part of the British countryside as the fox, its fellow in skulduggery, or indeed as the hares and rabbits which form its natural quarry. These puppies now chewing my trouser-legs and doing their utmost to remove my shoe-laces were, judging from their un-deniably villainous appearance, lurchers of a sort. The mother, now roaring appalling threats from behind the door of the shed, was, according to her owner, 'one of the very best'. She was the result of a union between a Welsh sheep dog/grey-hound dam and an old-fashioned fighting bull-terrier sire. The father of the pups was a well-bred whippet, although, looking at the little devils, I could hardly believe that their ancestry contained anything so refined.

After a certain amount of polite, if non-committal, con-versation, I turned and walked back to the gate. These puppies did not look as if they were likely to turn into the sort of dog which I had in mind. My ideal had always been the slender, rough-coated, deerhound-like type, of the sort one used to see trotting, tireless, self-sufficient and furtive as a shadow, in close attendance on the Romany's horse-drawn vardo.

I paused at the gate and glanced back. Rolling in my wake with jaunty, nautical gait, tongue lolling, and grinning confidentially up at me, waddled the undoubted pick of the litter, the only one out of the six to whom I had given more than a cursory glance. At eight weeks old he was a minuscule edition of his mother, except that down the length of his wide, intelligent forehead ran a broad white blaze, which, together with his snowy white chest, gave him an air of debonair rakehellishness. Seeing me pause he planted himself directly in front of me and leered into my eyes, his face one huge, red-tongued, glistening-toothed smile of pure comradeship. Had he actually spoken he could hardly have said more clearly: 'Why don't you and I team up, mate? With my brains and your brute force we could conquer the world!'

I picked him up and held him close, warm and solid in my arms. He reached out, took my nose between his tiny puppy teeth and gave it a tweak, gentle but determined. I put him on the ground and looked at him again with a new, proprietary interest. After all, he was going to be my dog. As he sat there, his leer more conspiratorial than ever, I could visualize him wearing a cloth cap and muffler and waving a jemmy. With his air of burglarious raffishness I dubbed him 'Sykes' on the spot, spelt with a 'y' to give it at least a hint of style. £2 was a reasonable enough sum to have paid for such a lovable lump of incipient villainy. I tucked Sykes, who appeared to have had no doubt whatsoever as to the outcome of our meeting, under my pullover and we returned to the car. I have not, in the last eight years, had reason to regret my decision.

Neville's friends lived a few miles on the other side of Llandovery in a grey, tree-shrouded house not far from the banks of the River Towy. Hearing the car's approach, they gathered beneath the shelter of a great flat-topped cedar, like a deputation of colonial officials awaiting the arrival of a respected but unpredictable new Governor-General. I wrapped young Sykes, who had already taken command of the situation, amongst a pile of old and doggy blankets and lifted Random's box on to the gravel drive before the front door. I slid up the fastening, slipped in my hand and she, as always, immediately stepped aboard. I drew her out and, to the expected chorus of admiring exclamations, held her aloft with wings spread, poised on high like a Roman standard suddenly come to life.

Such are the machinations of fate that I had neglected to attach the leash to her jesses. As I stepped on to the lawn to greet my hosts I stumbled slightly. Random was thrown forward jerking the jesses from my fingers. With the momentum gained from my plunging descent, she shot from my fist, gave her usual three measured flaps to gain height, half circled over the lawn, and rose above the belt of trees that lined and hid the nearby river bank. There was, amongst the shaken audience, the proverbial deathly hush, until someone turned to me and inquired casually: 'Does she make a habit of this sort of thing?' Waiting for no further pleasantries, Neville and I leaped into the car and set off in the direction which we hoped, judging by the wind, she had taken.

Chapter Twenty

In the garden of 'Neuadd Lodge', close to the village of Cilycwm, Imogen Rees was happily busy, surrounded by her whippets. In a few days' time the Championship Show at Paignton in South Devon would be taking place and she had high hopes that her own small but much-loved kennels would be represented there. In one member of the kennels at least, she had much faith. The beautiful, little, blue bitch 'Imrose Blue Star', known to her friends and family as 'Lucy', had already done well in the show ring and at the annual Whippet Club show had carried off the award for the best black or blue in the show. Not a bad beginning, and if she kept her end up at Paignton, who knew what honours might lie ahead – even a Ticket at Crufts itself, with its hopes of canine immortality, was not beyond the bounds of possibility.

Thus Imogen mused, smiling to herself as she revelled in the sunlight that filtered through the foliage of the great horse-chestnut tree. Her work for the moment over, she was sitting in a deck chair for a blessed hour's relaxation before she had to tackle the onerous task of preparing the dogs' evening meal. The seven whippets, graceful as gazelles, were lying flat out in the short sweet-scented grass, drinking in the warmth through every fibre of their bodies, intoxicated with the life-giving heat as only whippets and lizards can be.

Drowsing in the flower-scented summer afternoon, lulled by the drone of insect wings busy among the honeysuckle, Imogen was barely conscious of a dark form like a low, swiftly-moving storm-cloud that passed over her head, momentarily obscuring the gap between the roof and the outflung branches of the horse-chestnut. She did not hear, seconds later, the dry rattle of talons on stone nor the brief rustle of folding wings. She didn't hear, but the whippets, their reflexes trigger-sharp, heard; and, without pausing to consider, shot across the garden to investigate and if necessary to join combat with the author of this strange, possibly dangerous, sound. What they saw awoke their deeply-imbued protective instincts, urging them on with volley after volley of high hysterical barking. Imogen, jolted from her pleasant stupor, hurried in pursuit, shouting to the dogs to stop their infernal racket.

Where animals were concerned, Imogen was almost without fear. She could break up the noisiest, bloodiest dog-fight with calm efficiency. But here was something quite outside her experience. Perched upon the loosely-built stone wall that formed a boundary to one side of the garden, dividing it from the track which led to the home farm perhaps half a mile above the lodge, was one of the largest and certainly the most fearsome bird she had ever come across. She needed little

ornithological knowledge to realize that it was an eagle, but how in heaven's name came it here? She knew the buzzards well enough — indeed, she saw them every day. She had also, on memorable occasions, seen the great, narrow-winged, hawk-tailed outline of the red kite, sailing with incomparable grace in airy circles high over the garden. She had even watched them, short-legged and maladroit, on the ground, searching unromantically for beetles amongst the dry remains of animal droppings in the fields close by. But, to the best of her knowledge, eagles were not part of the local fauna. None the less, as she well knew, in Wales anything could happen.

The dogs, crazy with excitement, were leaping and dodging, daring each other to lead the final assault, whilst the eagle, mouth slightly open and auburn hackles raised, was silently defying them to do their worst. Fearing for the safety, both of the dogs, who were rapidly becoming uncontrollable, and of the eagle, who she suspected must be overwhelmed by a concerted charge, she shouted to the pack both individually and as a gang. Finally asserting herself, she managed to shepherd the unwilling dogs, still hurling abuse, back into the house.

We knew, at least we hoped, that Random would not go far. As we raced along the lanes my eyes were ever on the horizon, alert for any signs of mobbing crows or tail-flirting machine-gun-voiced magpies who might give some clue as to which way the wanderer had passed. We paused to listen for sounds of corvine indignation but apart from the distant peaceful cawing of a party of rooks, who with their young were trailing out to feed on the pastures, all was silent.

I searched the skyline. Every buzzard, every heron, became momentarily an eagle, raising, then shattering, my hopes. Where could she have gone? Surely she couldn't have swung southwards and be heading towards Emsworth, two hundred miles away? Hope hovered on the brink of despair. I was prepared for her to take to the wild — indeed, only a few hours before I had expected and resigned myself to just that eventuality – but she had returned and I had believed the bond between us to have been strengthened by the experience. This was totally unexpected. By my own clottish clumsiness I had startled her. She had taken off in thickly wooded country where every wood, every valley, appeared exactly like the last. She might well become totally lost.

Something caught my attention. I grabbed Neville's binoculars, nearly decapitating him in my hurry to transfer them from his neck to my own. On the distant horizon, poised above the point where earth and sky meet, Random was sweeping low over the craggy hillside. For a second I held her in my field of vision before she dipped below the earth's rim and was gone. Only a second – but it was enough. Size can be deceptive, but no mere buzzard was capable of that single-minded, deliberate, onward drive; slow-seeming perhaps, but none the less placing the eagle amongst the fastest things that fly. Random was heading northwards. She would surely take stand before long, if only to consider the situation and decide upon her next move. We raced in her wake, cursing the fact that we were earthbound and longing for just one more confirmatory sighting.

Beyond the village of Cilycwm a small group of obvious English visitors, field-glassed, camera'd and tweeded, were gathered on a rocky mound. Bird-watchers without a doubt! This was confirmed by their ties, which seemed to be liberally sprinkled with replicas of some sort of wader (of a species probably unknown to science and certainly to me). They were all peering at the empty sky and gesticulating with shooting-sticks, Thermos flasks and anything that they could lay their hands on. Something sensational, it seemed, had passed that way only moments before our arrival. 'Excuse me,' I began, 'but I don't suppose you've seen a . . .' 'The golden eagle? Well, it flew over our heads just a short while ago. It was going that-a-way.' (Pointing in the direction which we thought she must have taken.) 'I hope you won't disturb it. They are very rare in this part of the world, you know.'

215

Not wishing to enter into a discussion on the British distribution of *aquila chrysaetos* we thanked them and sped on our way. Something seemed to guide us, to lead us forward. I felt a sudden surge of hope. Round every corner I expected to see once more that familiar form surging into view or perched, motionless as a piece of statuary, on some commanding look-out post. Pausing where a dusty, pot-holed drive led upwards to an unseen farm below the summit of Craig Las I heard at last the sound I had been waiting for. The snarling, nasal challenge of carrion crows came from somewhere out of sight. Following the sound I saw a gang of these unlovely predators rising like bits of black, wind-blown paper to turn and stoop again and again with deceptive agility at something down below, something that was stirring them to raucous outbursts of fury.

Bumping and jolting over the spring-destroying pot-holes we passed the tiny white-painted lodge, the summit of the giant horse-chestnut now crowned with a cursing phalanx of crows joined by a posse of grey-naped jackdaws, ever on the alert for mischief. We drew up beside the mossy stone wall, disturbing a pair of magpies that were prancing about ready to turn any other creature's misfortune to their own advantage. I got a toe-hold on the crumbling wall, grabbed a handful of tough-stemmed willow-herb that was growing, apparently rootless, from the stones, and heaved myself up. From the centre of the garden arose what might long before have been an air-raid shelter, now covered by a flower-bright rockery. Clasping the largest rock with, as Tennyson put it, crooked hands, eyeing the mob of now silent hoodlums with supreme contempt, sat Random.

I hoisted myself over and walked up to her, doing my best to appear calm, casual, and in full command of my emotions. She looked at me, put her head on one side, and gave her curiously musical double-chirrup of welcome, exactly as if she were saying: 'Well, so here you are. You've taken your time, haven't you?' (I have always had the somewhat uncomfortable

216

feeling that if I chose to make the effort I could fly as well as she can, and it is simply lack of enterprise that keeps me grounded.) In my excitement I had left both my glove and leash in the car. I had, however, a rabbit's hind leg in the falconer's bag slung from my waist. I gave her this and she took it in a rather off-hand manner, examined it and began absent-mindedly to feed.

I went back to the lodge and hammered on the door. Amidst a chorus of barking the door opened and the lady of the lodge stood before me. 'I am so sorry to disturb you,' I said, 'but there's an eagle in your garden.' 'That,' replied Imogen Rees, 'is a fact of which I am only too well aware.' This was an unpromising beginning, but things began to improve when Imogen asked us in to tea, in the course of which we found that we had much in common, particularly a deep, life-long involvement with dogs and their welfare. Whilst Random finished her rabbit-leg, once more ensconced upon her block in the back of the car, we talked merrily on, discussing every aspect of dogs, dog breeding and dog shows.

Imogen's favourite was the whippet, a breed of which up to then I had had little experience, although I had always admired their symmetry, their speed, and above all their curiously medieval appearance, like a lion rampant, a look which they share with all the other breeds of 'gaze' hounds. Watching a litter of twelve-week-old puppies at play, I decided that I would like to know more about these endearing little dogs, that I might even like to own one. A sudden, searing pain shot through my left leg just above the ankle, a pain that might have been caused by the envenomed bite of a tarantula or possibly a black widow spider.

It wasn't a black widow that had bitten me – it was a black whippet. 'Laguna la Rousse' (Black Magic to her owner or friends if she had any) had emerged yawning and stretching from her basket in the kitchen. Hearing strange voices she had decided to investigate. Magic disliked strangers, particularly strange men. Travelling at the regulation whippet

217

speed (approximately two hundred yards in twelve seconds) she had in passing removed a sizeable slice from the calf of my leg, and when I reached the ground again she was vanishing into the shrubbery. I decided that I did not like whippets and that nothing on the face of this earth would persuade me to own one. (Magic and I have, since that unfortunate introduction, become almost inseparable.)

Imogen brought some of her dogs over to the farm next day for a walk in the forest and to exercise the local population of conies, which despite the loathsome wave of myxomatosis had now returned in force. I was boasting happily that in just two days I had made my dogs practically sheep-proof. Imogen had her doubts but was too polite to air them. Round the first bend we ran into a small flock cropping the grassy verge. At our sudden appearance the sheep, instead of standing still and allowing us to pass, trundled off in mindless panic. As one dog the combined pack took after them and, despite my oaths and imprecations, cornered a straggler against the forestry fence.

Bracken, the biggest and most powerful of them all, was dancing about like a Zulu warrior, uncertain whether or not to press home the attack. In desperation and with an agility that surprised me I brought her down with a tackle worthy of a British Lion and, luckily having the lead in my pocket, fastened her to the fence. Then I turned to mete out chastisement to the others, who were by now almost berserk with hunting ardour. Together Imogen and I managed at last to assert our authority and bring the chastened mob to heel. The sheep, heavily fleeced, escaped unscathed, but it is not one of my happier memories. My arrogance took a dive to zero, which on reflection was probably no bad thing.

We had just restored some sort of order and were slinking homewards when round the corner, mounted on a sturdy Welsh cob, came the farmer himself. He looked at us, he looked at the penitent pack. 'You've got a lot of dogs there, bach,' he remarked – an observation that was all too painfully self-evident. Looking closer, he went on, 'I hope those greyhounds are under control with you.' Resisting, but only just, an impulse to drag the fellow from the saddle, I replied as nonchalantly as possible: 'These dogs have been used to sheep all their lives. They wouldn't give one a second glance – they're only interested in hares and rabbits. Do you think that I'd take them about like this if there was any chance of them interfering with stock?' He looked at me, mumbled something in Welsh that sounded uncomplimentary and trotted on round the bend.

There is nothing like an adventure shared, particularly a hazardous experience, to bring two people with common interests even closer together. Two days later, while standing beneath a sparrow-hawk's nest on a soaking wet Cardiganshire hillside at midnight, Imogen and I became engaged, to a background accompaniment of snapping twigs and Welsh oaths, as the owner of the land on which the nest-tree stood blundered about in the upper branches. He was trying to rescue two well-grown eyases, which had been orphaned by someone who, in defiance of the law, had shot both parents because, so he alleged, they had been taking his young pigeons. Peter Panting slithered to the ground. From his coat pocket he produced two white, fluffy, animated puff-balls. On hearing our news, he solemnly pressed one of the eyases into my hand. 'I cannot at this moment think of a more suitable engagement present,' he remarked. In the dripping darkness, with the water pouring down our necks and the wind shrieking through the upper branches of the larch tree, echoed by the distant call of a hunting tawny owl, we made a vow. When we were married we would start a Bird Garden, a place where orphaned or injured birds of all kinds would be welcomed, restored if possible to health and ultimately released to live the life for which they were intended.

All this was very much in the future. Our immediate concern was with the little sparrow-hawk – 'Tal-y-fan' – who had come to us so suddenly out of the night to share our lives at so pertinent a moment. I tucked her, warm and trusting, beneath my pullover and together Imogen and I turned and climbed the long, heavily-wooded slope towards the glowing lights of the old Welsh mansion standing solid and timeless as the mountain behind it. Its door was half open in welcome, as if to tell us that here, in this wild and lovely country, our future lay, and that here we would find the encouragement to carry out the work for which, had we but known it, we had so long been destined.

And so, warmed and re-invigorated by numerous cups of strong coffee well laced with whisky, we returned to the shooting brake. As I opened the door I glanced at Random, who, of course, had come with us on this, for us, auspicious journey. She was standing utterly relaxed upon her block, one foot half hidden as is her wont amongst her breast feathers. She gave one of her companionable, slightly interrogatory, musical hooting calls as I stroked her broad back and tickled her under the chin. I switched on the light in the roof of the car and she looked me full in the eyes with her level searching gaze. It seemed so right that she should be there, because, after all, she, had she but known it, was in a way responsible for the whole affair. One thing was certain – whatever happened in the future, she, the presiding genius, would always be part of our lives for as long as she wished to stay with us. After all, her life and ours were much too closely knit for it to be otherwise.

Postscript

It is now sixteen years since Random came into my life. Sixteen years may not be an eternity but it is none the less nearly a quarter of a lifetime. To be owned by a golden eagle for so many years is an experience not enjoyed by many. I am proud and grateful for the privilege of living on intimate terms with such a fantastic bird, of knowing her in all her many moods and of being accepted by her as almost an equal. The golden eagle is not the rarest bird in the world; indeed, it is not even the rarest eagle. None the less, it is as magnificent as any – the eagle, *par excellence*, of fact and fiction, with mental and physical qualities entirely its own. It is, no doubt, for this reason that it has been able to hold on and to exist where other less adaptable eagles, such as the white-tailed sea eagle, have been exterminated.

In the past two decades I have lived with and trained a number of eagle species, including the comical, rather prehistoric-looking bateleur, with its ridiculously short tail, immense wing-span and incomparable flight; the sluggish, almost vulturine, carrion-eating tawny eagle; the lethal, rapacious, but, in my experience virtually untrainable, Bonelli's eagle; and finally, the golden eagle's first cousin, the handsome, aggressive imperial eagle. Of these species, not one can compare in any way – looks, temperament or performance – with Random – which, of course, is one reason why she is still with me whilst all the others, with the exception of the imperial, have either returned to the wild or been found homes elsewhere. Our long partnership has not been entirely without incident, and by no means all those incidents have been pleasant. At one time, her unreasonable, single-minded hatred of Imogen, inspired by jealousy, caused considerable un-happiness and embarrassment to all of us. When she killed my newly-trained and beautiful Swedish goshawk, Eira, after un-tying three different sorts of knot and hopping over a fifteen foot wire-netting fence, I almost banished her for ever. Only the thought of the emptiness her absence would have left behind in my life, coupled with the belief, that finding herself abandoned miles from home, she would take the law into her own talons and do something short, sharp and probably bloody, prevented me from doing this. The fact that, given the size and strength, Eira would have been equally willing to murder Random was poor consolation.

All this happened eight years ago. Since then she has travelled hundreds of miles with us, has been the star attraction at lectures, television appearances and all sorts of country events, and has done much by her personal appearances to forward the work of conservation and, above all, to show how poor a place the world would be without her and her kind to add their own particular glamour to the few wild places left to us. Always ready with the unusual, she astonished me and all who knew her by suddenly and without warning producing two eggs this year, the first appropriately on Easter Day.

I have recently obtained a prospective mate for her, a young and exceedingly handsome imperial eagle from Pakistan. Whilst these birds are of different species, they are, in my belief, closely enough allied to be able to produce a family if their initial mutual suspicion can be overcome. Whether, in fact, they will settle down to a life of wedded bliss, only time and nature can tell. Should they do so and be successful in rearing hybrid young 'imperial golden' eagles, it would prob-ably be the first time in history that such a union had taken place – but then, after all, Random, like Napoleon, has always had a natural bent for doing the unexpected.